Artificial Intelligence

Rethinking Human Consciousness and Agency in the Age of Machinic Simulation

Sergio Torres-Martínez

2025

MÍMIR EDITORS, Medellín, Colombia

By Sergio Torres-Martínez

ISBN: 9798230639725

Typesetting and art by

Sergio Torres-Martínez

Contents

Executive Summary

The unprecedented rise of generative artificial intelligence (GenAI), exemplified by systems such as OpenAI's ChatGPT and Anthropic's Claude, has introduced a transformative paradigm shift in technology's role in human society. This book critically examines the epistemological, philosophical, and practical implications of this shift, emphasizing the profound challenges posed by GenAI to our understanding of intelligence, agency, and consciousness.

Central to the analysis is the distinction between human cognition and artificial systems, particularly large language models (LLMs). While LLMs demonstrate remarkable proficiency in linguistic generation and problem-solving, their abilities remain rooted in statistical pattern recognition, devoid of the embodied intentionality, phenomenological experience, and biological imperatives that define human cognition. This book contends that reducing intelligence and creativity to computational processes oversimplifies the richness of human experience and risks fostering misleading narratives about the nature of AI.

Key findings and themes

1. ***The epistemological Fracture***: GenAI technologies have shifted the landscape of knowledge creation, challenging traditional linguistic and cognitive theories. The book highlights the limitations of LLMs in modeling human cognition due to their lack of embodiment, intentionality, and subjectivity.

2. ***Consciousness and Simulation***: The simulation of human-like thought and language by AI does not equate to genuine consciousness. Building on phenomenological and cognitive science perspectives, the book explores the philosophical

inadequacy of equating machine behavior with human awareness.

3. Human Agency vs. Machine Agency: Human agency is deeply tied to biological embodiment and survival-driven intentionality. AI systems, by contrast, lack these imperatives, functioning as tools for optimization rather than autonomous agents. The book advocates for recognizing these differences to avoid anthropomorphizing machines.

4. Cultural and Ethical Implications: The proliferation of GenAI raises critical questions about creativity, authorship, and the human role in cultural production. The book warns against over-reliance on algorithmic systems, which risk diminishing human creativity and eroding the phenomenological depth of cultural narratives.

5. Design and Semiotics: In examining the design of robot systems, particularly biomorphic ones, the book critiques their inability to replicate the semiotic richness of biological systems. It highlights the danger of viewing life through mechanistic lenses that undervalue the complexity of biological diversity.

Through five meticulously argued chapters, the book dismantles claims that GenAI systems are viable models of human cognition or capable of replacing human conceptualizers. Instead, it advocates for an interdisciplinary framework that situates AI within a broader understanding of humanity's unique cognitive and existential capacities.

Key Contributions

- *Critical Framework:* Provides a nuanced critique of AI narratives, emphasizing the need for a philosophical and cognitive reevaluation of human-machine distinctions.

- *Embodied Cognition*: Positions human embodiment as central to understanding intelligence, creativity, and consciousness, offering a counterpoint to disembodied computational models.

- *Ethical Guidance:* Proposes strategies to responsibly integrate AI technologies without undermining human agency or cultural richness.

This book is essential for academics, technologists, and policymakers seeking to make sense of the complex interplay between human cognition and machine intelligence in an era of transformative technological change. It also invites readers to reconsider what it means to be human in a world increasingly influenced by artificial agents.

Provenance of the chapters

The chapters in this book are grounded in a rigorous body of research, combining peer-reviewed publications with original, unpublished contributions. Chapter 1, published in *Biosemiotics*, offers a critical analysis of large language models through the lens of embodied cognition. Chapter 2, from *Language and Semiotic Studies*, challenges established linguistic theories with a predictive human model of language. Chapter 3, forthcoming in *AI & Society*, expands on these ideas by addressing the persistence of organic consciousness in the face of artificial intelligence. Chapter 4 introduces groundbreaking, unpublished work on the role of generative AI in shaping literary meaning. Finally, Chapter 5, published in *Biosystems*, examines embodied essentialism in robotic animal design. Together, these chapters provide a comprehensive and authoritative investigation into the complex interplay between human cognition, language, and AI.

Chapter 1: Torres-Martinez, S. (2024) Embodied human language models vs. large language models, or why Artificial Intelligence cannot explain the modal be able to. *Biosemiotics 17*,185–209. https:// doi. org/ 10. 1007/ s12304- 024- 09553-2

Chapter 2: Torres-Martinez S (2024d) A predictive human model of language challenges traditional views in linguistics and pretrained transformer research. *Language and Semiotic Studies, 10*(4), 562-592. https:// doi. org/ 10. 1515/ lass-2024- 0018

Chapter 3: Torres-Martinez S (2025 forthcoming). A predictive human model of language challenges traditional views in linguistics and pretrained transformer research. Dehumanizing the human, humanizing the machine:

Organic consciousness as a hallmark of the persistence of *the Human* against the backdrop of artificial intelligence. AI & SOCIETY. https://doi.org/10.1007/s00146-024-02165-x

Chapter 4 (Unpublished): Credence, attribution and creativity in the construction of literary meaning with generative artificial intelligence

Chapter 5: *Biosystems*: Torres-Martínez, Sergio. 2024b. Embodied essentialism in the reconstruction of *the animal sign* in robot animal design. *Biosystems* 238. 105178. https://doi.org/10.1016/j.biosystems.2024.105178

INTRODUCTION

The rapid ascent of generative artificial intelligence (GenAI), embodied by systems such as OpenAI's ChatGPT and Anthropic's Claude, has catalyzed a paradigm shift in our relationship with technology. These advanced systems, often heralded as potential replacements for human agents in the creation of culture and knowledge, have ignited a complex discourse oscillating between speculative optimism and existential dread. The implications of these technologies extend far beyond operational utility, revealing a profound epistemological rupture whose consequences are as yet inadequately understood.

Central to this disruption is the emergence of technologies that appear to emulate human agentivity, challenging deeply ingrained assumptions about decision-making, creativity, and the nature of intelligence itself. This rupture is not limited to practical considerations but ventures into the philosophical—forcing us to confront fundamental questions about the nature of humanity and its delineation from the machinic. What does it mean to possess agency, to create meaning, or to engage with the world as an intentional subject? In what ways do these technologies reflect, extend, or undermine our understanding of cognition and consciousness?

This volume advocates for a critical but measured perspective on artificial intelligence, eschewing hyperbolic claims of technological transcendence or obsolescence in favor of a nuanced examination of the boundaries between simulation and genuine experience. Building on seminal works such as Searle's Chinese Room argument (1980) and Levy's exploration of machine consciousness (2009), the book asserts that the question of whether machines can achieve consciousness is, at its core, misplaced. Rather, it redirects the discourse toward the phenomenological conditions of human consciousness itself—an experience intimately tied to biological embodiment, sensory integration, and intentional agency.

Generative AI systems, despite their remarkable computational capabilities, remain fundamentally constrained by their disembodiment. Lacking the biological imperatives that ground human cognition—such as survival-driven intentionality and embodied interaction with the environment—these systems operate as sophisticated simulacra rather than genuine cognitive agents. Nevertheless, the allure of attributing human-like qualities to these machines has fostered a proliferation of reductive narratives that equate linguistic fluency with conceptual understanding and information processing with creativity.

This book challenges such reductionism by addressing the critical distinction between human and artificial modes of cognition. Through a multidisciplinary approach, it interrogates the semiotic, philosophical, and cognitive frameworks underpinning claims about machine intelligence. Key themes include the epistemological limits of large language models (LLMs), the ontological differences between computational and biological forms of agency, and the ethical implications of ascribing consciousness to disembodied systems.

Organized into five chapters, this book provides an in-depth analysis of how GenAI is reshaping the landscape of knowledge and creativity. The essays compiled here reflect the outcomes of my research over the past year. While some overlap exists among the essays—particularly in my critique of posthuman perspectives on AI technologies and the philosophical missteps made by scientists and researchers concerning the nature of human and machinic consciousness—each piece offers unique insights into these pressing issues.

Chapter 1 critiques the linguistic assumptions underlying LLMs, contrasting their statistical generative processes with the embodied, agentive nature of human language. Chapter 2 develops a predictive model of human cognition rooted in biological adaptation, emphasizing the limitations of pre-trained transformers as tools for modeling reality. Chapter 3 explores the philosophical implications of posthumanist narratives, contesting claims of human obsolescence and reaffirming the centrality of embodiment to consciousness. Chapter 4 addresses the role of AI in literary and cultural production, warning against the erosion of human creativity through over-reliance on algorithmic processes. Finally, Chapter 5 examines the semiotics of machine-animal interactions, highlighting the risks of anthropocentric biases in robot design and their implications for the conceptualization of life itself.

Ultimately, this work is not merely an inquiry into artificial intelligence but a rigorous investigation into the essence of humanity. By interrogating the conceptual boundaries between the human and the machinic, it seeks to illuminate the enduring significance of embodied agency, intentionality, and the phenomenological richness of human experience in an increasingly technologized world.

CHAPTER 1

Embodied human language models vs. Large Language Models, or why Artificial Intelligence cannot explain the modal *be able to*

Abstract

This paper explores the challenges posed by the rapid advancement of artificial intelligence, specifically Large Language Models (LLMs). I show that traditional linguistic theories and corpus studies are being outpaced by LLMs' computational sophistication and low perplexity levels. In order to address these challenges, I suggest a focus on language as a cognitive tool shaped by embodied-environmental imperatives in the context of *Agentive Cognitive Construction Grammar*. To that end, I introduce an *Embodied Human Language Model* (EHLM), inspired by Active Inference research, as a promising alternative that integrates sensory input, embodied representations, and adaptive strategies for contextualized analysis and conceptual utility maximization. By incorporating Active Inference, which sees perception as inferring the world's state from sensory data, the findings reveal that the characterization of the English modal *be able to*, as a triadic construction encoding biological intelligent agency, introduces a more plausible theoretical basis for the positing of linguistic constructions. This emphasizes the crucial role of embodied human language models in the comprehension of how humans construct preferred futures through language.

Keywords: Active inference; Agentive Cognitive Construction Grammar; Embodied Human Language Model; Large Language Models; perplexity; triadic constructions.

1 Introduction: The biosemiotic stance

People like to say that we're insane
But AI will reward us when it reigns
Pledge allegiance to the world's most powerful computer
Simulation: it's the future

Grimes, *We Appreciate Power*, from the album *Miss Anthropocene* 2020

Artificial intelligence has been conceptualized in terms of functions geared by mathematical abstraction. However, it is increasingly accepted that any model of artificial agency requires taking into consideration multimodal interaction in connection to a niche. As a result, real-world constraints and noise (Sun et al., 2023) are required in order to understand the complex relations between machinic agents, environment, and living organisms. One important strand of biosemiotic research, to which this paper seeks to add (e.g. Romanini & Lacková,, 2023; Sharov, 2018; Sharov & Tønnessen, 2021), has concentrated on defining the role of agency in the construction of meaning.

> These include humans, animals, plants, single-cell organisms, individual cells in multicellular organisms, families, colonies, populations, ecological consortia, human communities, businesses and nations, autonomous mechanisms, robots, functional protein complexes in cells, viruses, and modified or

> engineered organisms. (Sharov & Tønnessen, 2021, p. 4).

This definition of agency has a lot to say for it. Among other things, human agency is not construed as the outcome of a mechanistic process of comparing inner data cycles against incoming information encoded in symbolic strings. Crucially, the biosemiotic conceptualization of human inference and linguistic ability brings into focus the uniqueness of biology in shaping our grasp of the world, since "[a]gents have a *different ontological status* than physical objects because they do not just exist, but exist for their internal purpose" (Sharov, 2018, p. 213). Moreover, by stressing the idea that agents are ecological actants, some implications of artificial agency modeling become evident, for example, anthropocentric views that reduce animacy (in the construction of robot animals) to the development of bioinspired objects lacking biological history and priority in an ecosystem. This is the case of quadruped robots (adaptable to uneven environments), robot fish, flapping-wing robots, etc., that, though equipped "with decision capabilities to react to external changes and the learning ability to adapt to new environments" (Wang et al., 2021, p. 1), do not possess self-referential symbolic autonomy. It needs to be said that animal robots, that is, bio-hybrid organisms (Romano 2018) whose body is intervened by means of biological control technologies such as implanted electrodes in the animal's brain (see, Peng et al., 2019; Zhou at al., 2022), are also defined as goal-constrained objects designed "in line with the 3R principle 'reduce, replace and refine' addressed to limit the use of experimental animals in research" (Romano, 2018).

The *robot animal sign* embodies an essentialist, solipsistic view of living organisms, defining their ontological status by function. Function is seen as a mechanistic reaction-action assigning specific roles in an Umwelt shaped by competition and environmental adjustment mimicking human cognition. A robot animal sign (RAS) represents a bioinspired technological myth that reinterprets animals' biomorphic traits for the completion of human-defined tasks. Therefore, the AI-powered RAS aims to optimize, minimizing trial-error learning, rather than enhancing ecologically oriented fitness through adaptive behavior. This conceptualization views RAS as performers lacking adaptive morphology, unable to gauge their impact on the environment. The challenge in designing ecological RAS lies in defining a machinic Self-symbol capable of measuring its actions as an extension of a meaningful environment.

A criterion used to define such continuity involves the capacity of living organisms to employ specific affordances, as exemplified in the ability to exhibit compensatory behaviors for injuries, which goes beyond the construction of "esthetic believability" (Leone, 2023, p. 4). The unifying semiotic mechanism consists of a bodily sign integrating trial-and-error learning algorithms, drawing inspiration from animal injury compensation. This approach aims to develop robots that can adapt and introduce behaviors addressing potential damage (Cully et al., 2015). This behavior also suggests the potential of non-animal species to adapt to and learn from their environment. Notably, iconic proximity (verisimilitude) holds little significance in this process. The ultimate implication is that the creation of AI-powered robot animal signs should not merely reflect a human inclination toward producing "fakes" (Leone, 2023, p. 6), a characterization that overlooks the necessity of a kind of "philosophical awareness" (Coghlan & Parker, p. 24) regarding the moral status of non-human entities, whether organic or artificial.

Instead, it should signify the emergence of non-human forms of intelligence capable of generating machinic semioses (see Torres-Martínez, 2021). Consequently, artificial robot signs are not Cartesian simulations of sentience—defined as the extended embodied response to external stimuli and their further coupling with models of internal states to attain semiotic equilibrium—but potentially adaptive entities. In the near future, they may possess a consciousness of internal states meaningful to them, assisting them in optimizing strategies for niche construction, which refers to the activities undertaken by organisms to "modify their surroundings" (Odling-Smee et al., 2003)

1.1 Human linguistic representation and large language models: The biosemiotic insight

When we compare human linguistic representation with the processing abilities of de-agentic pre-trained transformers such as ChatGPT, we come to see that these linguistic machines are not true cognizers. They are opaque performative pattern finders that lack experiential accuracy and hence tend to "hallucinate", as their abilities to distinguish reality from fiction are still very poor. Thus, although AI levels of performance in sentence parsing are currently used to compare (Dehouck & Denis, 2023), and even explain human linguistic processing (Huber et al., 2023), human agent representation is grounded in finding solutions and decision-making.

Large Language Models (LLMs) like ChatGPT excel in linguistic tasks—generating language, understanding input, conversing, classifying text—by analyzing vast text data and recognizing patterns. However, they fall short in defining certain language aspects, such as communication purpose, event boundaries, and emotional dimensions conveyed by linguistic choices. While LLMs offer insights and generate text with specific choices, nuances and subtleties in diverse language contexts still remain beyond their grasp.

Importantly, while some LLMs utilize *in-context learning* (ICL) to integrate syntax and grammar, which involves example demonstrations without updating any parameters, the ability to learn emergent syntax and grammar arises from a fusion of stored constructional semantics and learned patterns through techniques such as *few-shot prompting*. This approach has gained popularity in the development of LLMs such as GPT-4, which can be adapted to a wide range of tasks with minimal training data, based on the idea that quantitative changes can lead to qualitative behavior change (Anderson, 1972). In particular, ChatGPT's language generating capabilities are grounded in a deep learning model that employs multi-layer transformer architecture to generate language (Bomasani et al., 2022).

1.2 Perplexing *perplexity*, or why corpus studies need a linguistic theory

As noted, the advent of LLMs has introduced an unprecedented amount of fluid and reflowable linguistic data capable of emulating human linguistic communication. This has challenged many of the assumptions of theoretical linguistics (Linzen & Baroni, 2021), which puts pressure on the practicality of collecting and using large corpora and, of course, on the ontological status (for the modeling of linguistic theories) of the results obtained through the analysis of corpus-identified patterns. By definition, corpus results are metaphors constructed on machinic processing abstractions to reconcile frequency and probability measures (e.g., Divjak, 2023). Once converted to electronic data, linguistic samples (signs pointing to agentive human behavior) become affected by the perplexity values (a measure for the processing efficiency of language models) of the language model used to extract/generate them (they become flattened symbols for machinic extraction and quantification). In other words, human language usage is translated into **probability patterns** to be interpreted by all sorts of language models. In this context, *Language models* (LMs, see Devlin et al., 2019) are statistical theories that learn the probability distribution over sequences of words in a language. ChatGPT and n-grams are two types of language models that can be used to generate language. For example, the probability of generating a sequence of words w1, w2, ..., wn using an LM can be calculated as follows:

$$P(w1, w2, ..., wn) = P(w1) * P(w2 \mid w1) * P(w3 \mid w1, w2) * ... * P(wn \mid w1, w2, ..., wn\text{-}1)$$

where P(w1) is the probability of the first word, and P(wi | w1, w2, ..., wi-1) is the conditional probability of the i-th word given the previous i-1 words.

Similarly, the probability of generating a sequence of words using an n-gram model can be calculated using the following formalism:

$$P(w1, w2, ..., wn) = P(w1, w2, ..., wn\text{-}N+1) * P(wn\text{-}N+2 \mid w2, ..., wn\text{-}N+1) * P(wn\text{-}N+3 \mid w3, ..., wn\text{-}N+1) * ... * P(wn \mid wn\text{-}N+1, ..., wn\text{-}1)$$

where N is the order of the n-gram model, and P(w1, w2, ..., wn-N+1) is the joint probability of the first N-1 words.

1.3 A machinic linguistic ontology?

The question of whether experimentally collected data possess an ontological status, that is, whether they exist as objective referents in the world, or are merely instances of large perplexity values, is dramatically expressed by the language generation capabilities of ChatGPT. In a recent study, Winter et al. (2023) tested the preference for additive-rather-than-subtractive language in ChatGPT-3. According to the authors, the human cognitive preference for additive over subtractive language is also present in the pre-trained transformer:

> (...) the English language reflects a bias toward addition over subtraction in language statistics, as assessed via word frequency statistics, word order preferences in binomial expressions, contextual emotional valence and regular emotional valence, and the distributional semantics of "change" and "improve" verbs. These biases also carry over into the large language model GPT-3. Indeed, the quote that we started this paper with, which GPT-3 generated in response to the prompt word "add" is a positive word, beautifully reflects the extent to which this state-of-the-art natural language processing tool follows the "more is better" heuristic. (Winter et al., 2023, p. 20).

It is evident that equating ChatGPT's "additive bias" with corpus-identified human samples reduces data interpretation to flattened perplexity values lacking true ontological status for the identified "linguistic object" (e.g., Woodin et al., 2023). In particular, the object [ChatGPT-3 + additive bias] results merely from a flattened corpus/transformer perplexity value. Since ChatGPT cannot add pragmatic content beyond the prompt — that is, a flexible common ground of experience for the construction of human-chatbot interaction is missing (its bias being a product of the programming policies of the company developing these AI technologies)—the ontological status of Winter et al.'s study is compromised by statistical interpretation. Consequently, the transformer's "bias" is simply an adaptive response to specific user prompts through the attention mechanism in the transformer network.

In the next sections, I outline the nature of triadic constructions in the context of an *Embodied Human Language Model* (EHLM, see Section 3). Among other things, the EHLM introduces a mathematical reading of the mind-language interface (for the comprehension of human language learning and processing) that may help explain why the human mind and large language models create language in such different ways.

2 Language is experience, not a coarse simulation of it

The core argument presented here challenges the notion that syntax and lexis are merely hierarchical levels of symbolic expression organized within networks dictated by context-driven rules.[1] Among other things, the present model targets

1 Wittgenstein's *Philosophical Investigations* (1984 [1953]) provided baffling ideas regarding the nature of language and how normative use is subject to the rules of specific language games and forms of life (a way of doing and living within a community). I argue that Wittgenstein's idea that rule interpretation is akin to rule postulation is false, since the reasons to do something according to a rule are not dependent on goals existing prior to the accomplishment of a "given task". The idea I am advancing here, then, is that an expected consequence of endorsing the rule-following paradox is that any institutional use of language, in the context of a form of life or language game, is always open to a a deep-seated misconception in cognitive linguistics and construction grammar (e.g., Goldberg, 1995, 2006), namely that an interpretation of grammar is in itself a semiotic system subservient to a non-biological meaning-making mechanism (e.g., Jackendoff, 2012, p. 155). By highlighting survival imperatives as language's ultimate goal, it becomes evident that the need of agent conceptualizers to update beliefs over time (see also Pietarinen & Beni, 2021, p. 507) offers a more plausible understanding of language as a tool rather than an isolated system. Traditional construction grammar thus faces a challenge, as many constructions resist neat classification as stored parts or manipulable schemas, which contradicts the conventional usage-based view of language as a tool for cognitive reconstruction.

---

hypothesis-driven reconfiguration of the content of the rule due to the intentional character of actions. More concretely put, the content of rules is defined, not by *epistemic blindness,* and the systematic adherence to conventions (as Wittgenstein suggested), but the predictive character of human cognition. My interpretation posits that both the content of the rule and one's grasp of it fall under the broader category of reasons, thereby eluding a reduction of human action to a strictly "correct" or "incorrect" application of a rule. The reason is that rules *do not contain* behavior (as some wrongly claim, e.g., Ez-zizi et al., 2023), as reasons do not contain a norm to act in ways that privilege correctness over goals and purpose. Perhaps better put, agents engaging in a social use of language will always have the possibility to modify a given course of action in order to accrue epistemic capital. It is interesting in this connection to note that the possibility of accepting the components of a rule also means that my engagement with the rule itself is defined by *self-interest*, that is, in Parfit's words (1984, p. 10), that "[f]or each person, there is one supremely rational ultimate aim: that his life go, for him, as well as possible." Indeed, the idea that linguistic interpretations are rules for grasping the world according to a fixed form of life contradicts the cognitive imperative pushing us to strive for a form of preferred future in line with an individual's own intentions and goals. In other words, "closed thoughts" are simply not to be had (see Berto, 2022, pp. 3-4).

Understanding language as part of experience thus emphasizes agents as open biological systems reconciling experience with purpose:

1. Language as a Predictive Tool

- Language serves as a cognitive mechanism that helps intelligent agents anticipate and respond to their environment

- Through language, we create mental models that allow us to simulate potential outcomes and plan appropriate actions

- This predictive capacity extends beyond immediate physical responses to include social interactions and abstract problem-solving

- Language helps us organize experience into patterns that guide future behavior and understanding

2. Embodied Traces in Language

- Our linguistic expressions retain imprints of physical and sensory experiences

- Abstract concepts are often grounded in bodily metaphors (e.g., "grasping" an idea, "weighing" options)

- These embodied traces reflect how our cognitive systems evolved from basic sensorimotor processes

- The way we structure and use language reflects our physical experience of existing in the world

3. Direct and Ecological Experience in Language

- Language emerges from our immediate, unmediated interaction with the environment

- Our brain doesn't add artificial layers of representation

to experience - rather, meaning arises directly from our ecological engagement

- This suggests that language is deeply rooted in our practical interaction with the world

- The structure of language reflects the structure of our lived experience

4. Language as Event Reconstruction

- We use language to recreate and communicate experiences based on our bodily interactions with the world

- This reconstruction process draws on stored patterns of sensorimotor experience

- Language allows us to share these reconstructed experiences with others

- The ability to reconstruct events through language enables learning from past experiences

5. Language as a Constructed Semiotic System

- Language develops as an organized system of signs that includes:

  - Indexical elements (pointing directly to things)

  - Iconic elements (resembling what they represent)

  - Symbolic elements (conventional representations)

- This system emerges through collective human activity rather than being pre-generated

- The construction of language reflects both individual and social needs for communication

6. Non-Arbitrary Nature of Language

- Rather than being purely arbitrary, language patterns reflect shared cognitive tendencies among humans

- These patterns emerge from common human experiences and cognitive architecture

- Universal features in languages suggest underlying cognitive constraints

- Cultural and linguistic diversity exists within these cognitive parameters

7. Language as a Functional Assemblage

- Language is a dynamic system assembled from various cognitive, social, and environmental components

- It evolves to meet communicative needs while maintaining internal coherence

- The structure of language reflects its functional role in human activity

- Different aspects of language (phonology, syntax, semantics) work together as an integrated system

2.1 Why *Agentive Cognitive Construction Grammar* is more about cognition than constructions?

Agentive Cognitive Construction Grammar (AgCCxG, Torres-Martínez, 2023a,b, 2024) is a theory of mind and language that emphasizes how intention and purpose shape linguistic selections. It is cognitive, because linguistic selections arise from the need of biological open systems to balance a symbolic Self with the environment's dynamics. In contrast to usage-based approaches (e.g., Goldberg, 2006), AgCCxG relies on a strong theory of signs that defines human representation as continuous hybridization of sign systems in order to reduce uncertainty. In this view, language is construed as an embodied semiotic tool (see Torres-Martínez 2018), and not an assemblage of sequentially-arranged symbols connected within the system through a hierarchy of "paradigmatic associations" (Hilpert et al., 2023, p. 255; Schmid, 2020). Clearly, this characterization, rooted in semantic inheritance through degrees of family resemblance, favors a metaphor of computational efficiency that overlooks the cognitive dimension of language (see Silvennoinen, 2023). This perspective holds the belief that theoretical soundness relies on processing modeling, automatically assuming psychological validity.

2.2 AgCCxG and modality

As noted, this perspective becomes particularly compelling in the case of modality, defined herein, not simply as "the linguistic expression of possibility and necessity" (Depraetere et al., 2023, p. 9), but as the agentive embodied reconstruction of reality and stance through effect-oriented action planning (see Torres-Martínez, 2023a,b). This means, among other things that "what is possible and necessary" is precisely the sum of our experience as projected onto an entropic future that needs to be negotiated. The role of language in this regard is to define the extent to which specific mental states require being adjusted to particular events in order to reduce uncertainty. This definition benefits from a perspective on modality as a semiotic continuum where modal words and expressions are not detached from their intentional component, even in the context of second language acquisition (e.g., Grabar et al., 2023). Furthermore, they are not restricted to the identification of linguistic routines and formulas governed by intersubjective variation (e.g., Hufel & Schmid, 2023).

The implication is that the *intentional component of language* becomes an overarching category in which both volition and epistemic stance are deemed to reflect the biological imperative of living systems to engage in both boundary preservation and the updating of system-environment relations. This puts pressure on the reduction of modal meaning to single verb semantics, pragmatic saturation (Leclercq & Depraetere, 2021), or unmotivated chunks extracted from corpora, since, as we have seen, the intentional component of language requires speakers to combine perceptual experience and event-related action.

The claim is that current psycholinguistic cognitive approaches fail to accurately capture human representations through language (Jiang & Nekrasova, 2007; Nekrasova, 2009). Therefore, an alternative to traditional data mining involves incorporating strong embodied theories to propose hypotheses for linguistic research. As I have demonstrated elsewhere (Torres-Martínez, 2018, 2023a), the identification of agentive language profiles in large corpora relies on adhering to relevant theoretical premises.

For instance, defining constructions as triads (rather than dyads) of form-function and intelligent agency[2] allows a better understanding of modals' semantics in specific contexts beyond coarse collocational preferences in a corpus (see Torres-Martínez, 2018). Additionally, when constructional attachment tendencies are accurately identified, these constructions are characterized not as frequency-based strings (e.g., Hilpert, 2013), or corpus-

2 *Intelligent Agency* refers to actions in which agent-conceptualizers engage in a specific behavior during an event that is the product of both perceptual, hypothesis-driven states, and *intentional doxic (objectifying) lived experiences* (see Husserl, 1989, p. 5), and that are defined by the agent's intention to attain a goal.

identified modal words (Cappelle et al., 2023), but instances of experience with the world.

3. A human language generation model

The main thesis of this section is that the role of language, as part of a biological system, is to create a link between the organism's need to negotiate *uncertainty*, as a means to maintain *homeostatic* (inner-body) integrity, and our intended, goal-oriented action on the world.

A cautionary note is needed at this point. I do not intend to introduce an axiomatizable super theory of language aiming at a rigorous word-world alignment. Rather, it is intended to provide a multi-layered reconstruction of propositions internally connected through scalar amplification. As Wilson (2022, p. XIV) puts it, "If condition S becomes true upon the lower scale length ΔL, then condition S^* will become true on the higher scale length ΔH."

Indeed, as it will become evident in the next sections, the intrinsic value of a proposition across the conceptual layers in the theory implies a revision of the whole theory at each step of the process. Another point worth mentioning is the role of conventionality in the construction of meaning through language, since "(…) our use of language conforms to regularities-and that is all" (Lewis, 2002, p. 2). What is open to discussion is the nature of those regularities. Therefore, the main thesis of this section is that humans are open biological systems that need to mobilize a battery of "cost free or low-cost signals at equilibrium" (Barbieri, 2007, p. 301) in order to engage in agentive behaviours guided by *intention*. Intention is the result of being immersed in a sometimes hostile environment where survival imperatives shape linguistic selections. This points to a very important fact about human language usage, namely that it is not purely symbolic.

Under this view, a human language generation model, the *Embodied Human Language Model* (heretofore EHLM) can be posited using three levels of *linguistic reconstruction*: 1) The triadic-construction formula, 2) the constructional attachment formula, and 3) the embodied conceptualization formula.

3.1 The triadic-construction formula

The structure of triadic linguistic constructions points to the existence of a *Bodily Self* (agent, that is, A), whose action is expanded to the realm of unintentional sign-making. It follows that the structure of utterances, and what we make of them in terms of meaning construction, is not the result of a reasoned activity encoded in our minds thanks to "our being able to take a reflective stance towards utterances" (Jary, 2022, p. 2). The purpose of any proposition is not, then, to match some "propositional slot" in the receiver's head, but to extend experience beyond the conceptualizer's mind. Cognizers are thus in the position to use language in ways that expand the life of symbols to other sign forms. Language is thus defined by an extended form of semiosis axed in three semiotic trajectories "which enables us to produce different signifying itineraries with the same objects that function as interpreted signs and interpretant signs" (Petrilli & Ponzio, 2015, p. 57): iconic (imagistic) links provide a pre-cognitive association between concepts and referent objects in terms of form and function. During their passing from qualities (icons) to indices (signs conveying information in terms of contiguity and reaction), referent objects display relational properties. The definition of a triadic construction is this:

> Linguistic constructions are triads of *form, function, and intelligent agency* that emerge through the combination of specific constructional attachment patterns. These patterns are embodied constructional arrangements that organize both high-level elements (abstract syntactic patterns) and low-level constructions, such as words, morphemes, affixes, etc., in networks of interconstructional relations motivated by agentive selections.

The formula for the triadic construction can be represented as follows:

$$\mathbf{LC} = \{F, Fn, A\}$$

Where:

- *LC* refers to the triadic linguistic construction

- *F* represents the form of the construction, which can be high-level (abstract syntactic patterns) or low-level (words, morphemes, affixes, etc.).

- *Fn* represents the function of the construction, which can be semantic, pragmatic, or both. • *A* represents the intelligent agency involved in the construction, which can be the speaker, the listener, or both.

These three layers of linguistic reconstruction are organized in networks of interconstructional relations through CAPs that are motivated by agentive selections. These patterns are influenced by the goals, intentions, and preferences of the speaker and/or listener and can be used to communicate meaning and achieve communicative goals in a variety of contexts. The formula for Triadic Constructions (TC) integrating CAPs can be expressed as:

$$TC = A + CAP + (I + In + S)$$

Where:

- TC represents the Triadic Constructions
- A represents the bodily self or biological agent
- CAP represents the Constructional Attachment Patterns
- I represents Iconic links
- In represents Indexical links
- S represents Symbolic links

The above equation illustrates how *Triadic Constructions* emerge from the combination of the *Bodily Self,* or agent, specific CAPs, and semiotic links between concepts and referent objects, which are organized in three types of relations: iconic, indexical, and symbolic. The bodily Self, or agent, selects specific CAP arrangements that organize high-level and low-level constructions in networks of interconstructional relations. The semiotic links between concepts and referent objects provide the necessary form-function pairing that confers agentive properties to the Triadic Construction. Iconic links create pre-cognitive associations between concepts and referent objects in terms of form and function, indexical links provide information about contiguity and reaction among referent objects, and symbolic links connect concepts, form-functional relations with referents in the world.

3.2 Triadic constructions and Active Inference

The tripartite structure of constructions, inspired by *Active Inference* (AIF), a process theory applied in diverse disciplines (Friston, 2005, 2010; Clark, 2013), posits that the brain minimizes the discrepancy between its internal model and sensory inputs. AIF, prevalent in cognitive neuroscience and machine learning, asserts that the brain actively reduces uncertainty through Bayesian inference. Linguistic behavior is shaped by recurrent hypothesis semiotic models, referred to as *Constructional Attachment Patterns* (CAPs, Torres-Martínez 2018, 2021, 2023a), serving as templates for selecting specific constructional associations. These CAPs guide speaker selections based on "agentive imperatives", rather than statistical ones, thereby facilitating manageable categories in communication. Notably, this communication type does not adhere to a polar framework, which means that purpose extends beyond the definition of statefulness and statelessness during communication (Ludlow & Živanović, 2022, p.166). Agents are characterized by a type of bodily awareness that can be conceptualized as,

> *the representation of the Self in the world through the conjunction of perceptual states, conceptualizations thereof and the ensuing projection of a set of beliefs onto specific events in a context of use.*

Self-representation is theoretically explained through bodily-system integration (BSI) mechanisms, which are central for constructing bodily awareness and shaping our perception and conceptualization of the world (Nowakowski, 2017). The embodied dimension of cognition resides not merely on the involvement of specific body parts in events but on integrating species-specific bodily memory and online systemic adaptation through prediction-error minimization

(Torres-Martínez, 2023b). These reenactment principles capture the process of embodied reality reconstruction through language. The cognitive dimension of language isn't defined as the result of a computational problem-solving process embedded in the sensorimotor organization but is inseparable from the broader streams of meaning assignation responsible for purposeful information weighing mechanisms. Unlike *Eco-Cognitive computationalism* (Magnani, 2018), the present theory emphasizes the pragmatic dimension of embodied cognition, defining the role of agents within a system of relations not solely measurable as goal-directed statistical selections.

The belief in the truth of a proposition thus relies on the consolidation of an experiential fact via a consistent, biologically-driven modeling of concepts for uncertainty reduction. The question of a priori knowledge is not addressed by invoking absolute meaning, but by confronting facts subject to perceptual filtering (Boghossian, 2020).

3.2 Triadic Constructions + CAPs

In addition to the TC formula, a TC = CAPs formula can be posited that includes a variable R representing the *agentive properties* given to conventionalized form-function pairs:

$$TC = A + CAPs + (I + IX + S) + R$$

where:

- TC represents triadic linguistic constructions.
- A represents the bodily self or agent.
- CAPs represents specific Constructional

Attachment Patterns that organize high-level and low-level constructions.

- I represents the iconic links that provide a pre-cognitive association between concepts and referent objects.

- IX represents the indexical links that convey information in terms of contiguity and reaction.

- S represents the symbolic links that connect concepts and referents in the world.

- R represents the agentive properties given to conventionalized form-function pairs.

This equation implies that triadic linguistic constructions arise from the combination of four elements: the bodily self or agent (A), specific *CAPs* semiotic links between concepts and referent objects (I, IX, and S), and agentive properties (R) given to conventionalized form-function pairs. The agentive properties reflect the goals, intentions, and preferences of the speaker and/or listener, and play a significant role in organizing and interpreting linguistic constructions. The active inference model can be integrated with the Triadic Constructions + CAPs model by incorporating the agent's beliefs about the environment and its internal states into the TC = CAPs formula. As a result, I use the agent's beliefs to update the various components of the formula and guide action selection, thus:

$$TC = A(q) + CAPs(q) + (I + IX + S) + R(q)$$

where:

- TC represents triadic linguistic constructions.

- A(q) represents the bodily self or agent, with q representing the agent's beliefs about its bodily self or agentive properties.

- CAPs(q) represents specific Constructional Attachment Patterns that organize high-level and low-level constructions, with q representing the agent's beliefs about these patterns.

- I represents the iconic links that provide a pre-cognitive association between concepts and referent objects.

- IX represents the indexical links that convey information in terms of contiguity and reaction.

- S represents the symbolic links that connect concepts and referents in the world.

- R(q) represents the agentive properties given to conventionalized form-function pairs, with q representing the agent's beliefs about these properties.

To integrate Active Inference into the model, belief updating equations from active inference can be used that update the agent's beliefs concerning bodily self, specific CAPs, and agentive properties. Sensory input on body position updates beliefs about the bodily Self, while past experiences shape beliefs about agentive properties. Action selection is guided by the active inference action selection equation, thereby minimizing future surprise through updated beliefs about the environment and internal states.

The combination of the Triadic Constructions + CAPs model with active inference offers a framework elucidating how agent beliefs and actions, influenced by environmental interactions and internal states, contribute to linguistic construction emergence.

So, the EHLM predicts that human speaker intuition cannot be captured by smart learning systems (LLMs) from raw data only. This is because semantic and grammatical "latency" are not directly embedded in the human mind. Instead, this information is encoded in the *events* that provide a suitable context for language use. These events possess identifiable affordances[3] and contribute to the reaffirmation of the Self as a cognitive agent through the interplay of our perceptual Self and our conceptualizing Self.

This presents an insurmountable obstacle for machine learning, since open biological systems are constantly shaped by *core Umwelt* (Tønnessen, 2015) variables that are not available to LLMs. Thus, the construction of possible worlds has a strong embodied component that is represented by the preference of speakers to express (non)-factual reasoning through specific CAPS such as argument structure.

4. Applying the EHLM to the study of English modals

The previous characterization of modals as triadic constructions that reflect patterns of energy minimization through uncertainty reduction opposes some of the analyses provided by constructions grammarians. In particular,

3 An affordance (Gibson, 1966, 1977) is a potentiality that can be used by an agent in a given way according to its phylogenetic configuration. Objects can thus be used in manners defined by criteria such as weight, size, form, etc., that are perceptually accessible, which results in the emergence of specific action profiles.

Leclercq's and Depraetere's 2021 study (henceforth L&D) which splits modal meaning into either wide/narrow, or subject internal/external reconstruction:

1)

Cracks *can* appear overnight

According to L&D (2021, p. 6), the root-possibility proposition in (1) is characterized by a wide-scope modality, suggesting the postulation of a theory. However, the possibility of cracks appearing overnight is *a*-theoretical. To be more precise, there can be no wide scope (or, at least, not in the absolute sense predicted by L&D, see Section 4), as the agent conceptualizer is merely invoking shared world knowledge to mobilize their beliefs about a state of affairs. Similarly, the concept of a *source of modality* being "subject-internal/internal" fails to account for the fact that event reconstruction needs to be negotiated based on shared world knowledge, and re-enactment is a co-construction grounded in specific affordances.

Therefore, the proposition "I can touch my nose with the tip of my tongue" (Leclercq & Depraetere, 2021, p. 6) is not "internal" simply because the tongue emerges from inside someone's mouth. Instead, this case reveals that (i) others have tongues, creating a link between this act and the experiences of others, and (ii) my ability to stick out my tongue in this particular manner is an affordance of my bodily configuration, perceived by others as part of an event and not as an isolated instance.

Moreover, the notion of a "barrier to actualization" as a feature of root-permission modality is unnecessary as a conceptual device for classifying modal words. Event actualization results from complex affordance mappings and belief-structure actualization/reaffirmation, extending beyond mere permission by an agent. In the context of the proposition "You can just call me Billy" (Leclercq & Depraetere, 2021, p. 6), the idea that Billy becomes a "barrier to actualization" is incorrect. On closer inspection, it is Billy's belief structure (rather than his "being Billy") that creates conditions for extending the affordances of the event beyond formal interaction. *Conventional interaction* (not Billy) acts as an underlying, albeit weak, barrier to actualization, manipulable by agents to suit their needs. Clearly, the concept of permissibility is not effectively communicated by the combination of wide-scope modality and a "potential-barrier" criterion.

2)

We hope that all the Bangladeshi refugees *can* be repatriated in mid-August 2015

As is evident in (2), there is not a wide type of modality scoping over the entire proposition, since no theory is being postulated. Moreover, the alleged barrier to actualization is dissolved by predictive inference. Thus, the prediction that the refugees can be repatriated at some point is overridden by the conceptualizer's undisclosed world knowledge (supported by the mapping of previously experienced event's affordances including, for example, bureaucratic or political variables) whereby the stated deadline will not be met, or, worse, that the refugees will not be repatriated at all.

On this reading, English modal constructions reflect forms of intelligent agency, which means that perception and cognition influence each other back and forth or, as we have seen, become overridden by habitual perceptual routines that create a context for predictive action.

The mathematical formula for the embodied model of English modal constructions can be represented as follows:

$$MC = (IA + L) * ECM * DE * (BP + AP)$$

where:

- MC represents modal certainty
- IA represents intelligent agency
- L represents language
- ECM represents embodied content in the form of stored, pre-semiotic cognitive routines
- DE represents direct and ecological embodied experience
- BP represents belief-driven models of the world
- AP represents absolute forms of actualization or event potentiality

The formula reveals that English modal constructions emerge from the interaction of intelligent agency and language, both shaped by embodied content in the form of stored, pre-semiotic cognitive routines. Embodied experience is direct and ecological, without additional representational layers. Modal space event reconstruction thus mirrors factual event reconstruction, involving belief-driven world models and absolute forms of actualization or potentiality. This process, featuring action planning and affordance mapping, prioritizes perceptual convention over judgment for a joint event reconstruction. Thus, three integrated agentive domains emerge:

1. The proper agentic domain where the agent actively seeks to accumulate epistemic capital.

2. A domain of belief where past experiences are applied to a local or new situation.

3. A domain of re-enactment where the affordances of an animated object are mapped and further incorporated into a potential embodied model of the world.

Consequently, this paper defines modality as

> *the agentic embodied reconstruction of reality and stance achieved through effect-oriented action planning and affordance mapping.*

The central concept proposed here is that any action is contextualized, integrated into an event, and executed by an entity or an animated object/concept. Particularly in the context of epistemic modality, it is the agent conceptualizer who shapes the proposition's boundaries through an embodied, belief-driven alignment with experience.

We can conclude, based on the formula presented, that L&D's analysis of modal constructions in examples 1-2 oversimplify the complex factors involved in the expression of modality. First, the formula takes into account five different factors:

1. The degree of internalization of the agent (IA),
2. The linguistic expression of modality (L),
3. The event contingency matrix (ECM),
4. The degree of distance from the event (DE), and

5. The balance of power between the agent and the subject (BP + AP).

The degree of internalization of the agent refers to the extent to which the agent is involved in the event, which could include physical or mental involvement. L&D's analysis of "subject-internal/internal" modality does not fully capture the complexity of this factor, as there may be cases where the agent's internalization is not the main source of modality expression. It is interesting in this connection to note that this observation challenges the necessity of introducing nuanced levels of agentivity, such as fuzzy characterizations ranging from agents in the broadest sense to individuals, as proposed by Flach et al. (2023, p. 152).

The linguistic expression of modality includes modal words and constructions, as well as other linguistic cues such as tense and aspect. While L&D analyze the scope of modality expressions as either "wide" or "narrow," this does not fully capture the range of linguistic devices that can convey modality.

The event contingency matrix refers to the likelihood of the event occurring, as well as the consequences of the event. L&D's analysis of the "barrier to actualization" as a feature of root-permission modality seems to overlook the fact that event actualization is influenced by a range of complex factors beyond the permission of an agent.

The degree of distance from the event refers to how close or remote the event is to the speaker. This factor can influence the expression of modality, as well as the likelihood of the event occurring.

Finally, the balance of power between the agent and the subject refers to the relative influence and control each has over the event. This factor can affect the expression of modality, as well as the likelihood of the event occurring.

Thus, the formula MC = (IA + L) * ECM * DE * (BP + AP) can be used to calculate the degree of modal certainty (MC) in a sentence or proposition. Here's a brief explanation of each variable:

- IA: The degree of internal access to the event by the speaker/hearer. This refers to the speaker's/hearer's direct experience or knowledge of the event being described.

- L: The degree of linguistic evidence that supports the event or state of affairs. This refers to the linguistic cues present in the sentence or proposition that signal the likelihood of the event.

- ECM: The degree of epistemic control of the speaker over the event or state of affairs. This refers to the speaker's level of control or influence over the event being described.

- DE: The degree of discourse expectation that the event or state of affairs will occur. This refers to the extent to which the event is expected based on the context of the discourse.

- BP: The degree of background possibility of the event or state of affairs. This refers to the degree to which the event is possible based on general knowledge or background information.

- AP: The degree of actual possibility of the event or state of affairs. This refers to the degree to which the event is possible based on specific circumstances or context.

Now, let's explore the plausibility of the existence of absolute values of wide-scope modality. In (3), L&D assert that the possibility of cracks appearing overnight is theoretical. However, we can apply the formula to calculate the degree of modal certainty (MC) and argue that the sentence exhibits a low-rather-than-high degree of wide-scope modality. Let's assign the following values to the variables:

IA (Individual Attitudes): Low

L (Lexicalization): Medium

ECM (Event-Concept Mapping): Low

DE (Degree of Expectation): High

BP (Background Probability): Medium

AP (Agentive Probability): High

Using these values, we can calculate the modality certainty as follows:

$MC=(IA+L)\times ECM\times DE\times(BP+AP)$

$\mathrm{MC}=(0.2+0.5)\times 0.2\times 0.8\times(0.5+0.8)$

$MC=0.7\times 0.2\times 0.8\times 1.3$

$MC=0.182$

Similarly, in (2), L&D argue that there is no wide-scope modality in the sentence. However, we can use the formula to calculate the degree of modal certainty and argue that there is some degree of wide-scope modality. Let us assume the following values for the variables:

- IA: Low
- L: High
- ECM: Low
- DE: High
- BP: Medium
- AP: High

Using these values, we get:

MC= (IA + L) * ECM * DE * (BP + AP) MC = (0.2 + 0.8) * 0.2 * 0.8 * (0.5 + 0.8) MC = 1 * 0.2 * 0.8 * 1.3 MCs = 0.208

The calculations show that the degree of modal certainty is still relatively low, but there is some degree of uncertainty in the proposition. This suggests that there is some degree of wide-scope modality present in the sentence.

So we see that the formulas presented suggest that a more nuanced and multifactorial approach is necessary to fully grasp the expression of modality in language.

5. Embodied modal constructions

As already shown by Torres-Martínez (2018), native speakers of English tend to use specific argument structures to express possibility, which correspond to two main categories: *Modal Agentive Constructions* (MASCs) and *Modal Referential Constructions* (MRCs).

5.1 Modal Agentive Constructions (MASCs) involve actions

involving transitivity or literal motion across literal and modal mental spaces. In this sense, both root and epistemic meanings have an embodied component consisting of either transfer or intransitive forward momentum. MASCs use modal clusters, e.g., modal + full verb, frequently used to express epistemic obligation in the *Corpus of Contemporary American English* (COCA, Davies, 2008). The scope of modality in MASCs is narrow, and predicates always provide information about the type of agency expressed in the modal cluster (see Table 1.1).

(3)

"None of us ***needs to*** gamble, or drink alcohol, read e-mail, or compulsively check social networking feeds to survive."

(Daniel J. Levitin, *The Organized Mind: Thinking Straight in the Age of Information Overload,* 2015).

5.2 Modal Referential Constructions (MRCs) express *balance*, which is the possibility of adopting and maintaining a stable position in a direct relationship with gravity. This is often framed in terms of epistemic evaluations of reality. Two types of MRCs, *attributive* and *defining*, encode the human need to take a stance, evaluate qualities, or define states that enable us to adapt to events or contribute to the solution of problems. Epistemic constructions have evolved to help the conceptualizer define agentive roles, rather than semantic gradation. In MRCs, the agent-addressee becomes the source of modality, that is, the one who must restore balance. In contrast, the agent-conceptualizer evaluates the result of a current state of affairs, which is clearly out of his/her control.

(3) Attributive MRC

(5)

"Any man who achieves any stature through his own efforts accumulates enemies. Envy is the most common human failing. You ***must be familiar*** with this yourself."

(Glenn Cook, *The Tyranny of the Night*, 2010).

(6)

"Why me? What have I done to deserve an invitation to what ***has to be the biggest harebrained scheme*** of the century?" (Dirk Pitt 4, *Raise the Titanic*, 2004).

Table 1.1. Frequencies of modal constructions in the *Corpus of Contemporary American English*. (See High-res Figure at the end of this book).

Modal construction	TYPE	FREQ	PER MIL	FORM	FUNCTION
COULD BE (+NP)	Defining MRC	21,685	198.23	(Subj.$_{SOURCE}$) MODAL DO/BE NP	[X MODAL DO/BE Y_{DEF}]
COULD SEE	Transitive MASC	17,145	163.44	(Subj.$_{FORCE}$) MODAL V Obj.	[X MOD act on/experience [Y]$_{SCOPE}$
CAN GET	Caused-motion MASC	10,864	99.31	(Subj.$_{FORCE}$) MODAL V [Oblique Source] $_{SCOPE}$	[X MOD cause [Y to move from Z]$_{SCOPE}$
WILL GET	Intransitive-motion	9,509	86.93	(Subj.$_{FORCE}$) MODAL V [Oblique$_{PATH}$] $_{SCOPE}$	[X MOD move [Z_{PATH}]$_{SCOPE}$
WILL MAKE	Ditransitive MASC	4,232	38.43	Subj.$_{FORCE}$) MODAL V [Obj1 Obj2] $_{SCOPE}$	[X MOD cause [Y to receive Z]$_{SCOPE}$
WILL BE ABLE	Attributive MRC	3,039	27.78	Subj.$_{SOURCE}$) MODAL BE Adj (to phrase)	[X MODAL BE Y_{STATE}]

Modal construction	TYPE	FREQ	PER MIL	FORM	FUNCTION
COULD BE (+NP)	Defining MRC	21,685	198.23	(Subj.$_{SOURCE}$) MODAL DO/BE NP	[X MODAL DO/BE Y_{DEF}]
COULD SEE	Transitive MASC	17,145	163.44	(Subj.$_{FORCE}$) MODAL V Obj.	[X MOD act on/experience [Y]$_{SCOPE}$
CAN GET	Caused motion MASC	10,864	99.31	(Subj.$_{FORCE}$) MODAL V [Oblique Source] $_{SCOPE}$	[X MOD cause [Y to move from Z]$_{SCOPE}$
WILL GET	Intransitive-motion	9,509	86.93	(Subj.$_{FORCE}$) MODAL V [Oblique$_{PATH}$] $_{SCOPE}$	[X MOD move [Z_{PATH}]$_{SCOPE}$
WILL MAKE	Ditransitive MASC	4,232	38.43	Subj.$_{FORCE}$) MODAL V [Obj1 Obj2] SCOPE	[X MOD cause [Y to receive Z]$_{SCOPE}$
WILL BE ABLE	Attributive MRC	3,039	27.78	Subj.$_{SOURCE}$) MODAL BE Adj (to phrase)	[X MODAL BE Y_{STATE}]

6. Analyzing "*be able to*" using the EHLM

As already noted, the integration of the Triadic Constructions + CAPs model with Active inference provides a framework for understanding how the agent's beliefs and actions are shaped by their interactions with the environment and their own internal states, and how these beliefs and actions contribute to the emergence of linguistic constructions. This integration allows for a more sophisticated understanding of language processing than either model alone, as it incorporates both the agent's internal state and the external environment.

Thus, in the case of the modal "*be able to*", the TC=CAP formula can be applied as follows:

$$TC = A(q) + CAPs(q) + (I + IX + S) + R(q)$$

where:

A(q) represents the agent's beliefs about its *bodily Self* or agentive properties, which might include beliefs about its ability to perform certain actions or achieve certain goals.

CAPs(q) represents the agent's beliefs about specific constructional attachment patterns that organize high-level and low-level constructions, which might include beliefs about the likelihood of encountering certain constructions in the environment.

R(q) represents the agent's beliefs about the agentive properties bestowed upon conventionalized form-function pairings, which might include beliefs about how listeners are likely to interpret the modal "*be able to*".

By incorporating the agent's beliefs into the

TC=CAP formula, we can better understand how the agent processes the modal "be able to" and how s/he selects appropriate actions based on their beliefs and expectations (see Table 1.2). For example, if the agent believes that s/he has a high likelihood of success in achieving a particular goal, they may be more likely to use the modal *"be able to"* in their speech. Conversely, if the agent believes that s/he is unlikely to succeed in achieving a particular goal, they may be less likely to use the modal *"be able to"*.

We can see that the integration of the *Triadic Constructions + CAPs* model with the active inference model provides a more comprehensive framework for understanding language processing and how it is shaped by the agent's beliefs and interactions with the environment. This integration allows for a more nuanced understanding of constructions such as the modal "be able to" and how they are used in language.

Table 1.2. Frequency of *"be able to"* in the *Corpus of Contemporary American English.* (See High-res Figure at the end of this book).

SECTION	ALL	BLOG	WEB	TV/M	SPOK	FIC	MAG	NEWS	ACAD	1990-94	1995-99	2000-04	2005-09	2010-14	2015-19
FREQ	23145	0	0	0	23145	0	0	0	0	4053	3840	3979	3666	3909	3698
WORDS (M)	993	21.8	21.1	21.8	21.5	20.1	21.5	20.6	20.3	23.7	25.1	24.9	24.5	24.7	24.5
PER MIL	23.31	0	0	0	1,076.73	0	0	0	0	171.03	152.48	159.3	148.41	157.91	149.92

The TC formula (TC = A + CAPs + I + IX + S + R) can now be used to analyze the frequency and distribution of the triadic linguistic construction *"be able to"* in the different genres and time periods listed in Table 1.2.

A: The *bodily Self* or agent. In the case of "be able to", this refers to the subject of the sentence that is able to perform the action described by the verb. For example, in the sentence "I am able to run fast", the bodily self or agent is "I".

CAPs: Specific constructional attachment patterns that organize high-level and low-level constructions. In the case of "be able to", this refers to the grammatical structure of the construction, which typically consists of the auxiliary verb "be" and the infinitive "able to" followed by the main verb. For example, in the sentence "She will be able to swim soon", the CAPs of "be able to" are the combination of the auxiliary verb "will be" and the infinitive "able to swim".

I: ***Iconic links*** that provide a pre-cognitive association between concepts and referent objects. In the case of "be able to", this refers to the concept of ability or capacity to perform a certain action, which is associated with the referent object (the agent) that is able to perform the action.

IX: ***Indexical links*** convey information based on contiguity and reaction. In the case of *be able to*, this refers to the temporal and causal relationships between events described by the construction. For instance, in the sentence "He was not able to attend the meeting because of traffic," the link between "not able to attend" and "traffic" explains the reason for the subject's inability to attend the meeting.

S: ***Symbolic links*** connect concepts and referents in the real world. Regarding "be able to," this pertains to the symbolic representation of ability and capacity in language and its connection to the real-world agent. For example, "be able to" symbolically represents the idea of ability or capacity and links it to the agent who can perform the action.

R: ***Agentive properties*** are influenced by conventionalized form-function pairings. In the case of "be able to," this encompasses the social and cultural conventions that shape its use and interpretation in different contexts. For instance, the agentive properties of "be able to" may vary depending on the genre or time period in which it is used. The frequency analysis reveals that *"be able to"* is most common in spoken language (23,145 per million words) and least common in academic writing (0 per million words). This suggests that the agentive properties of *"be able to"* differ across genres, with spoken language emphasizing its ability to convey capacity or ability in real-time communication. Furthermore, Table 1.2 indicates that the frequency of "be able to" has remained relatively stable over time, with a slight increase in the 2015-2019 period compared to earlier periods. This suggests that its CAPs (I, IX, and S) have remained largely consistent, while its R has been influenced by evolving social and cultural conventions.

7. Conclusion

In this paper, I have argued that the introduction of a human language model based on genuine cognitive principles expands the possibilities for theoretical advancement and data analysis. Among other things, this challenges conventional methods of linguistic data interpretation relying on corpus-based fragment analysis (e.g., Hilpert & Perek, 2022).

The claim is that language arises from cognitive processes aimed at minimizing uncertainty, which offers a more nuanced framework that reduces the need for unnecessary classifications, taxonomies, and unmotivated prototype identification. One notable contribution of LLMs, as part of what Hoffmeyer (2008, p. 344) called *semiotic technologies*, is their ability to simplify language analysis. As machines effortlessly generate language without syntactic abstractions and rules, linguists should develop language models that explain the diversity of human language as a feature governed by cognitive imperatives for the construction of knowledge, that is, "the combination of qualia, objects and interpretants in semiosis" (Favareau, 2009, p. 712), rather than the interpretation of linguistic data.

References

Barbieri, M. (2007). *The codes of life: The rules of macroevolution*. London and New York: Springer.

Berto, F. (2022). *Topics of thought: the logic of knowledge, belief, imagination*. Oxford: Oxford University Press.

Boghossian, P., & Williamson, T. (2020). *Debating the A priori*. Oxford: Oxford University Press.

Bommasani, R., Hudson, D. A., Adeli, E., Altman, R., Arora, S., von Arx, S., ... Leskovec, J. (2022). On the opportunities and risks of foundation models. *arXiv preprint arXiv*:2108.07258.

Cappelle, B., & Depraetere, I. (2016). Short-circuited interpretations of modal verb constructions: Some evidence from The Simpsons. *Constructions and Frames, 8*(1), 7-39. https://doi.org/10.1075/cf.8.1.02cap.

Cappelle, B., Depraetere, I., & Lesuisse, M. (2019). The necessity modals have to, must, need to and should: Using n-grams to help identify common and distinct semantic and pragmatic aspects. *Constructions and Frames, 11*(2), 220-243. https://doi.org/10.1075/

cf.00028.cap.

Cappelle, B., De Cuypere, L., Depraetere, I., Grandin, C., & Leclercq, B. (2023). Possibility modals: Which conditions make them possible? In S. M. Fitzmaurice & B. Kortmann (Eds.), *Models of modals: from pragmatics and corpus linguistics to machine learning* (pp. 93-117). Berlin-Boston: De Gruyter Mouton.

Chomsky, N., Gallego, Á. J., & Ott, D. (2019). Generative grammar and the faculty of language: insights, questions, and challenges. *Catalan Journal of Linguistics, Special Issue*, 229-261.

Clark, A. (2013). Whatever next? Predictive brains, situated agents, and the future of cognitive science. *Behavioral and Brain Sciences, 36*(3), 181-204. https://doi.org/10.1017/S0140525X12000477.

Coghlan, S., & Parker, C. (2023). Harm to non-human animals from AI: A systematic account and framework. *Philosophy and Technology*, 36. https://doi.org/10.1007/s13347-023-00627-6

Davies, M. (2008). *The Corpus of Contemporary American English (COCA): 520 Million Words, 1990-Present.* Available online at http://corpus.byu.edu/coca/.

Devlin, J., Chang, M.-W., Lee, K., & Toutanova, K. (2019). BERT: Pre-training of deep bidirectional transformers for language understanding. In *Proceedings of the 2019 Conference of the North American Chapter of the Association for Computational Linguistics: Human Language Technologies, Vol. 1 (Long and Short Papers)* (pp. 4171–4186). Minneapolis, Minnesota: Association for Computational Linguistics.

Depraetere, I., Cappelle, B., & Hilpert, M. (2023). Introduction. In S. M. Fitzmaurice & B. Kortmann (Eds.), *Models of modals: from pragmatics and corpus linguistics to machine learning* (pp. 1-13). Berlin-Boston: De Gruyter Mouton.

Divjak, D., Romain, L., & Milin, P. (2023). From their point of view: the article category as a hierarchically structured referent tracking system. *Linguistics*. https://doi.org/10.1515/ling-2022-0186

Ez-zizi, A., Divjak, D., & Milin, P. (2023). Error-Correction Mechanisms in Language Learning: Modeling Individuals. *Language Learning*, 1-37. https://doi.org/10.1111/lang.12569

Favareau, D. (2009). Essential readings in biosemiotics: Anthology and commentary. London and New York: Springer.

Flach, S., Cappelle, B., & Hilpert, M. (2023). You *must/have to* choose: Experimenting with choices between near-synonymous modals. In S. M. Fitzmaurice & B. Kortmann (Eds.), *Models of modals: from pragmatics and corpus linguistics to machine learning* (pp. 149-176). Berlin-Boston: De Gruyter Mouton.

Gibson, J. J. (1966). *The senses considered as perceptual systems*. Boston: Houghton Mifflin.

Gibson, J. J. (1977). *The ecological approach to visual perception*. Boston: Houghton Mifflin.

Goldberg, A. E. (1995). *Constructions: A Construction Grammar Approach to Argument Structure*. Chicago: University of Chicago Press.

Goldberg, A. E. (2006). *Constructions at Work: The Nature of generalization in language*. Oxford: Oxford University Press.

Friston, K. (2005). A theory of cortical responses. *Philosophical Transactions of the Royal Society of London, 360*, 815–836. https://doi.org/10.1098/rstb.2005.1622.

Friston, K. (2009). The Free-Energy Principle: A rough guide to the brain? *Trends in Cognitive Science, 13*, 293–301. doi: 10.1016/j.tics.2009.04.005

Grabar, Hamon, T., & Leclercq, L. (2023). Modals as a predictive factor for L2 proficiency level. In S. M. Fitzmaurice & B. Kortmann (Eds.), *Models of modals: from pragmatics and corpus linguistics to machine learning* (pp. 199-224). Berlin-Boston: DeGruyter Mouton.

Hilpert, M. (2013). Die englischen Modalverben im Daumenkino: Zur dynamischen Visualisierung von Phänomenen des Sprachwandels. *Zeitschrift für Literaturwissenschaft und Linguistik, 42*, 67-82.

Hilpert, M., & Perek, F. (2022). You don't get to see that every day: On the development of permissive get. *Constructions and Frames, 14*(1), 13-40.https://doi.org/10.1075/cf.20011.hil.

Hilpert, M., & Flach, S. (2023). Modals in the network model of construction grammar. In S. M. Fitzmaurice & B. Kortmann (Eds.), *Models of modals: from pragmatics and corpus linguistics to machine learning* (pp. 254-269). Berlin-Boston: DeGruyter Mouton.

Hoffemeyer, J. (2008). *Biosemiotics: An examination*

into the signs of life and life of signs. Scranton and London: University of Scranton Press.

Huber, E., Sauppe, S., Isasi-Isasmendia, A., Bornkessel-Schlesewsky, I., Merlo, P., & Bickela, B.(2023). Surprisal from language models can predict ERPs in processing predicate-argument structures only if enriched by an Agent Preference principle. *Neurobiology of Language*. Advance publication. https://doi.org/10.1162/nol_a_00121

Hufeld, C., & Schmid, H.J. (2023). Does the intersubjectivity of modal verbs boost inter-individual differences? In S. M. Fitzmaurice & B. Kortmann (Eds.), *Models of modals: from pragmatics and corpus linguistics to machine learning* (pp. 177-196). Berlin-Boston: DeGruyter Mouton.

Jackendoff, R. (2012), *A user's guide to thought and meaning*. Oxford: Oxford University Press.

Jary, M. (2022). *Nothing is said: Utterance and interpretation*. Oxford: Oxford University Press.

Jiang, N., & Nekrasova, T. M. (2007). The processing of formulaic sequences by second language speakers. *The Modern Language Journal, 91*(3), 433–445. https://doi.org/10.1111/j.1540-4781.2007.00621.x.

Leclercq, B., & Depraetere, I. (2021). Making meaning with be able to: modality and actualisation. *English Language and Linguistics, 26*(1), 27-48. https://doi.org/10.1017/S1360674320000341.

Leclercq, B., & Depraetere, I. (2021). *Making meaning with be able to: modality and actualisation. English Language and Linguistics, 26(1), 27-48*. https://doi.org/10.1017/

S1360674320000489

Leone, M. (2023). The main tasks of a semiotics of artificial Intelligence. *Language and Semiotic Studies*. https://doi.org/10.1515/lass-2022-0006

Lewis, D. (2002). *Convention: A philosophical study*. Oxford: Blackwell Publishers.

Linzen, T., & Baroni, M. (2021). Syntactic Structure from Deep Learning. *Annual Review of Linguistics, 7*(1), 195–212. DOI: 10.1146/annurev-linguistics-011619-030303.

Ludlow, P., & Živanović, S. (2022). *Language, form, and logic: In pursuit of Natural Logic's Holy Grail.* Oxford: Oxford University Press.

Magnani, L. (2018). Eco-Cognitive Computationalism: from mimetic minds to morphology-based enhancement of mimetic bodies. *Entropy, 20,* 430. doi:10.3390/e20060430

Nekrasova, T. M. (2009). English L1 and L2 speakers' knowledge of lexical bundles. *Language Learning, 59*(3), 647–686. https://doi.org/10.1111/j.1467-9922.2009.00524.x.

Nowakowski, P. (2017). Bodily processing: the role of morphological computation. *Entropy, 19,* 295. doi:10.3390/e19070295

Odling-Smee, F. J., Laland, K. N., & Feldman, M. W. (2003). *Niche construction: The neglected process in evolution*. Princeton, MA: Princeton University Press.

Parfit, D. (1984) *Reasons and persons*. Clarendon Press: Oxford University Press.

Peng, Y., Wang, Z., Zhang, Q., Du, S., Zhao, Y., Yang, L., Liu, J., Cheng, Y., Wang, A., & Liu, Y. (2019). Basic research on wireless remote control rabbit animal robot movement. In H. Yu, J. Liu, L. Liu, Z. Ju, Y. Liu, & D. Zhou (Eds.), *Intelligent Robotics and Applications. ICIRA 2019. Lecture Notes in Computer Science, vol. 11741.* Springer, Cham. https://doi.org/10.1007/978-3-030-27532-7_4

Petrilli, S., & Ponzio, A. (2015). Language as primary modeling and natural languages: A biosemiotic perspective. In E. Velmezova, K. Kull, & S.J. Cowley (Eds.), *Biosemiotic perspectives on language and linguistics,* pp. 47-76). London and New York: Springer.

Piantadosi, S. (2023). Modern language models refute Chomsky's approach to language. *LingBuzz.* Retrieved from https://lingbuzz.net/lingbuzz/007180

Pietarinen, A-V, & Beni, M. (2021). Active Inference and abduction. *Biosemiotics,* 499-517. https://doi.org/10.1007/s12304-021-09432-0

Romanini, V, & Lacková, L. (2023). Morphoesthetics in artificial intelligence: proteins versus machines. *Transdisciplinary Journal of Linguistics.* doi : 10.53987/2178-5368-2023-12-08

Romano, D., Donati, E., Benelli, G., & Stefanini, C. (2018). A review on animal–robot interaction: From bio-hybrid organisms to mixed societies. *Biological Cybernetics.* https://doi.org/10.1007/s00422-018-0787-5

Sarker, I. H. (2021). Deep learning: A comprehensive overview on techniques, taxonomy, applications and research directions. *SN Computer Science, 2, Article*

420. https://doi.org/10.1007/s42979-021-00711-3

Schmid, H.-J. 2020. *The dynamics of the linguistic system. Usage, conventionalization, and entrenchment*. Oxford: Oxford University Press.

Sharov, A. (2018). Mind, agency, and biosemiotics. *Journal of Cognitive Science, 19* (2), 195-228.

Sharov, A. & Tønnessen, M. (2021). *Semiotic agency: Science beyond mechanism.* Cham: Springer Nature.

Silvennoinen, O. (2023). Is construction grammar cognitive? *Constructions, 15* (1). doi 10.24338/cons-544

Tønnessen, M. (2015). Introduction: The relevance of Uexküll's Umwelt theory today. In C. Brentari (ed.), *Jakob von Uexküll: The discovery of the Umwelt between Biosemiotics and theoretical biology*, pp. 1-19. London and New York: Springer.

Torres-Martínez, S. (2018). Constructions as triads of form, function and agency: An agentive cognitive construction grammar analysis of English modals. *Cognitive Semantic, 4*(1), 1–38.

Torres-Martínez, S. (2021). Complexes, rule-following, and language games: Wittgenstein's philosophical method and its relevance to semiotics. *Semiotica, 242*, 63-100.

Torres-Martínez, S. (2023a). A radical embodied characterization of German Modals. *Cognitive Semantics*, *9*(1), 132-168.

Torres-Martínez, S. (2023b). An Integrated Bayesian-Heuristic semiotic model for understanding human and SARS-CoV-2 representational structures. *Biosemiotics*.

https://doi.org/10.1007/s12304-023-09546-7

Torres-Martínez, S. (2024). *Agentive Cognitive Construction Grammar*: A predictive semiotic theory of mind and language. *Semiotica (Accepted manuscript).*

Wang, J., Chen, W., Xiao, X., Xu, Y., Li, C., Jia, X., & Meng, M. Q.-H. (2021). A survey of the development of biomimetic intelligence and robotics. *Biomimetic Intelligence and Robotics.* https://doi.org/10.1016/j.birob.2021.100001

Wilson, M. (2022). *Imitation of Rigor: An alternative history of analytic philosophy.* Oxford: Oxford University Press.

Winter, B., Fischer, M. H., Scheepers, C., & Myachykov, A. (2023). More is better: English language statistics are biased toward addition. *Cognitive Science, 47,* e13254. https://doi.org/10.1111/cogs.13254

Wittgenstein, L. (1984 [1922]). *Werkausgabe, Band 1: Tractatus Logico-Philosophicus/Tagebücher 1914–1916 /Philosophische Untersuchungen.* Frankfurt am Main: Suhrkamp Verlag.

Woodin, G., Winter, B., Littlemore, J., Perlman, M., & Grieve, J. (2023). Large-scale patterns of number use in spoken and written English. *Corpus Linguistics and Linguistic Theory.* https://doi.org/10.1515/cllt-2022-0082

Zhou, Z., Mei, H., Li, R., Wang, C., Fang, K., Wang, W., Tang, Y., & Dai, Z. (2022). Progresses of animal robots: A historical review and perspectiveness. *Heliyon,* 8, e11499. https://doi.org/10.1016/j.heliyon.2022.e11499

CHAPTER 2

A predictive human model of language challenges traditional views in linguistics and pre-trained transformer research

Abstract

This chapter introduces a theory of mind that positions language as a cognitive tool in its own right for the optimization of biological fitness. I argue that human language reconstruction of reality results from biological memory and adaptation to uncertain environmental conditions for the reaffirmation of the Self-as-symbol. I demonstrate that pre-trained language models, such as ChatGPT, lack embodied grounding, which compromises their ability to adequately model the world through language due to the absence of subjecthood and conscious states for event recognition and partition. At a deep level, I challenge the notion that the constitution of a semiotic Self relies on computational *reflection*, arguing against reducing human representation to data structures and emphasizing the importance of positing accurate models of human representation through language. This underscores the distinction between transformers as posthuman agents and humans as purposeful biological agents, which emphasizes the human capacity for purposeful biological adjustment and optimization. One of the main conclusions of this is that the capacity to integrate information does not amount to phenomenal consciousness as argued by *Information*

Integration Theory. The conclusion is that while language models exhibit superior computational capacity, they lack the real consciousness providing them with multiscalar experience anchored in the physical world, a characteristic of human cognition. However, the chapter anticipates the emergence of new *in-silico* conceptualizers capable of defining themselves as phenomenal agents with symbolic contours and specific goals.

Keywords: Active Inference, Agentive Cognitive Construction Grammar, ChatGPT, embodiment, essentialist concept formation, large language models

1. Introduction

In this chapter, I introduce a theory of mind that aims to provide a context for the analysis of language as a cognitive tool for the optimization of biological fitness. In doing so, I also explore the potential of construction grammars to become either theoretical contributors to, or ontological beneficiaries from the development of large-language-model developments. However, in contrast to traditional construction grammars, I argue that the human language reconstruction of reality is the result of both biological memory and adjustment to uncertain environmental conditions, and the need to define a preferred future in the reaffirmation of the *Self-as-symbol*.

Among other things, I show that pre-trained language generators are incapable to create a model of the world through language due to their lack of embodied grounding. As I shall indicate further below, autoregressive pre-trained transformers create models of the world that reflect an essentialist, single-layered reconstruction of reality that lacks embodied representation owing to an absence of subjecthood and conscious states for event recognition and partition. Importantly, though humans engage in symbolic essentialism through language (concepts about natural kinds, for example, are generally modeled on essential and/or intrinsic properties ascribed to an entity or object), our capacity to manipulate syntax and semantics well beyond the prediction of statistically-identified tokens, enables us to integrate, blend and overlap eventful experience to create suitable pictures of the world. More concretely put, while humans use their intuitive physics and biological memory to acquire concepts and express them through a vast amount of constructional combinations, *Large Language Models* (LLMs) mobilize large amounts of data to infer syntactic and semantic suitability as a means to increase efficient response rates to a prompt. Since human performance in communication is not defined by controlled task completion but survival, assertions that transformers "exhibit subtle and sophisticated pattern identification and inferencing" (Lappin, 2024, p. 11) conflate *performance* with intelligent agency, a hallmark of human language learning, usage, and entrenchment.

Importantly, empiricist readings of LLM performance are not alone in ascribing intelligent agency to AI-generated language. Indeed, amodal language processing theories of language have seen in LLM research a good means to tackle the undeniable paucity in the investigation of how human language is processed and conceptualized in the speaker's mind. These approaches come in a variety of forms depending on the definition of how language is produced and processed. What they have in common is the goal of creating mathematical theories that accurately capture the way meaning is *extracted* from symbols and further mapped onto internal representations and concepts possessed by speakers immersed in an ideal state of *unity* "often thought of as part of a descriptive theory of ideal rational agents, not of real agents" (Kindermann & Onofri 2021, p. 2).

It makes sense, then, to begin our investigation by pointing out that this unified representationalist assumption rests on the arbitrariness of the symbols used. In many ways, an undue distinction between the linguistic symbol and perception is created as a condition for *logical closure,* which contrasts with embodied theories that, in many cases, reject the idea that representations are internal and entirely amodal. Although symbolic models depart from different assumptions regarding the ways concepts are distributed and organized, whether they are local (Collins & Quillian, 1969), or distributed (Hinton et al. 1986), current symbolic models of cognition, such as *latent semantic analysis* (Landauer & Dumais, 1997), or *word2vec* (Mikolov, Chen, Corrado, & Dean, 2013) offer design rules for the manipulation of signs through algorithms that distribute words across semantic layers in "high-dimensional numerical vectors extracted from large amounts of natural-language data" (Günther et al., 2019, p. 1), following mathematical compression techniques for the construction of representations that work at the level of prediction and performance rates during specific tasks. Running across these distributional semantic models (Günther et al., 2019, p. 4) is the idea that language can be coded, encoded, and further transformed into retrievable patterns defined as language-as-performance instances capturing human linguistic decisions (thereby becoming "explanatory theories" Landauer & Dumais, 1997) within distributional vectors.

We can see the problem with the symbolic account manifest itself in many different ways outside the realm of *Good Old Fashioned Artificial Intelligence,* especially in the context of *generative artificial intelligence* (GenAI) and large language models (LLMs). As we will see in the next subsection, researchers coming from an empiricist and functionalist tradition have gone so far as to elevate LLM performance (that is, language-as-performance) to the status of "the best predictive models of human language representations at the resolution of data [neuroscientists] have access to" (Tuckute et al., 2024, p. 285), in a manner akin to the claim that symbolic distributional models can "take sensorimotor experience as context" (Günther et al., 2019, p. 15).

1.1. Humanism or posthumanism in large language model conceptualization

In a similar way, the notion of *language-as-performance* has been integrated into ethnographic (posthumanist) viewpoints seeking to ascribe an ontological status to pretrained transformers in the belief that these language generators are *authentic* language creators (Demuro & Gurney, 2024). According to this reading, such models actively contribute to the cultural construction of knowledge through unique reasoning processes that, nonetheless, are distinct from those of humans. This assertion stems from a rejection of a "universalist" definition of language, which, according to some authors, is often associated with a male-centered humanist tradition, and instead views language as a manifestation determined by the performance of communication. Consequently, pretrained language models are seen not just as performers of language, but as active participants in its creation, thereby shaping events where language unfolds through both performance and competence, which "relativize[s] the human by coupling it to some other order of being" (Clarke, 2008, p. 3). Accordingly, human language is depicted as a process of assembling symbols within a pre-established grid of semiotic relations, with speakers functioning as participants bound by the rules of extant signhood preservation practices within a communicative event. As a result, human language is portrayed as a residual aspect of interaction, lacking any inherent transcendental qualities ("humanness"), while positioning LLM language generation as an integral component of authentic "encounters" where participants, both human and non-human, share a sense of experiential adjustment (e.g. Demuro & Gurney, 2024; Dynel, 2023). This shift obliterates any claim to a construction of the Self-as-symbol through an emphasis on a redefinition of power relations that replace universalism with a focus on addressing gender inequality.

> My anti-humanism leads me to object to the unitary subject of Humanism, including its socialist variables, and to replace it with a more complex and relational subject framed by embodiment, sexuality, affectivity, empathy and desire as core qualities (Braidotti, 2013, p. 26).

However, since human agency involves purposeful, goal-oriented action aimed at both reducing uncertain states and constructing preferred models of interaction with the environment, human beings, as distinct biological systems, participate in a continuum of relations with the world and other species that goes beyond denaturalized claims to power redistribution. The primary contradiction in this posthumanist reasoning lies in the fact that even the most contingent set of relations is driven by biological imperatives that cannot be relativized through the postulation of indefinite intrinsic properties: "For posthuman theory, the subject is a transversal entity, fully immersed in and immanent to a network of non-human (animal, vegetable, viral) relations ." (Braidotti, 2013, p. 193).

Hence, if we were to accept that artificial language generators create culture simply by existing in the world through interactions with humans (Demuro & Gurney, 2024), we are also compelled to believe that humans, as open biological systems, depend on an infinite, regressive chain of decisions made at a molecular level by non-cognizant interpreters in order to justify their existence.

This form of semiotic atomism, as part of what some have called "new materialisms" (e.g. Coole and Frost 2010) assigns portions of purposeful animacy and cognition to fine-grained instances or states of a system's parts, under the belief that bottom-up self-organization cannot be explained by observing the system *as is*, but rather as it could be under different environmental conditions (e.g. Watson, 2024). This antihumanist stance is summarized by Pepperell's assertion (2003, p. 2) according to which "the balance of dominance between human and machine is slowly shifting". In other words, human uniqueness is being evaluated through an arbitrary and biased notion of efficiency (through information capacity and memory, e.g. Cantlon & Piantadosi, 2024) that renders human agency unnecessary and relativized as idiosyncratic manifestations of cognitive actants drowning in a sea of "material forces [that] manifest certain agentic capacities" (…) (Coole & Frost, 2010, p. 10) Inevitably, the blurring of the line between conscious (human?) and unconscious (artificial) states (subsumed to the rules of performance), embedded within unwarranted part-whole relations, fails to provide a coherent characterization of subjecthood and experience as core elements of organic agency:

> The *zoe*-centred embodied subject is shot through with relational linkages of the contaminating/viral kind which inter-connect it to a variety of others, starting from the environmental or eco-others and include the technological apparatus. (Braidotti, 2013, p. 193)

The obvious result of this reasoning is the postulation of political relativism "devoted to the ["critical"] analysis of actual conditions of existence and their inherent inequality" (Coole & Frost, 2010, p. 25). As we shall see, the postulation of humans, "a strange product of evolution that can perceive itself and its fellow humans as paradoxical" (Rosendahl, 2013, p. 224), as self-symbols, striving for biological priority and persistence, is a requirement for the definition of language as a cognitive tool that cannot be emulated by artificial systems of language generation engrained within a fuzzily and ideologically defined material environment in which "there is no definitive break between sentient and nonsentient entities or between material and spiritual phenomena" (Coole & Frost, 2010, p. 10) . Importantly, my use of "persistence" does not seek to position a code-driven idea of biological reality through heredity. As it will become evident in the next pages, my idea is, rather, anchored in the definition of signhood within a context of predictive semiosis, and not the passing of information through DNA-based heredity (Birch et al., 2020, p. 56).

1.2. Defining consciousness

Where is, then, consciousness to be found? According

to *Information Integration Theory* (IIT), the question of consciousness needs to be defined from a first-person perspective. In this vein, IIT considers two problems: 1) the extent to which a system can be considered as having conscious experience, and 2) the definition of the conditions that determine the type of consciousness a system has.

This phenomenological approach departs from the assumption that, for system to be conscious of itself and of what is going on around it, it must, first, be able to deal with large amounts of information (differentiation), and second be capable of integrating that information into a coherent semiotic picture of the world (integration). Moreover, the whole process is said to be private and occur in specific areas of the brain. The main weakness of this approach is that it cannot address some important question regarding the role and nature of consciousness, especially with regard to purposeful action. Whereas some could argue that a focus on a need to cope with uncertainty in an ever-changing environment is enough to account for the rise of different types of consciousness, it is a fact that living organisms are not isolated systems, but create complex networks of relations with other organisms, including those considered as exhibiting disparate levels of consciousness. In other words, consciousness is possible because it provides us with a sense of extended community that creates connections with others. Consciousness is thus constructed upon patterns of interaction associated with agentive perception that can be projected onto the environment and other entities. The implication is, of course, that the one condition of conscious states is that can be shared an enhanced by collective experience. This leads us to a second objection to ITT: its reductionism of conscious states to an impoverished version of the brain as a physical support of experience (a quantity).

The present proposal defines consciousness as a

combination of four main elements: *1)* The identification of objects in the world associated with events of perception (experience is intrinsic, though modeled on shared experiential structures in common with other organisms). *2)* The construction of pictures of the world through the integration of modal and amodal information processes in the brain (information is provided by different modalities and then organized in the brain in specific areas. The nature of experience is defined by degrees of perceptual access to *complexes* (composite concepts) and not by simulations of experience in the form of *"shadows of perception"*. *3)* The content of experiential states becomes integrated into the agentic projection of intentional states onto the world. *4)* The alignment of subjective conscious states with other conscious entities is facilitated by the construction of a model of the self both as individual and part of a system of agentive relations and hierarchies. This notion of bodily awareness, as an instance of agency, is indebted to the *free-energy principle* (Friston, 2009, 2010) whereby organisms need to keep a balance between environmental and inner-body information in order to avoid entropy. This defines the role of biological systems as predictive entities moving toward specific *attracting states*, that is, points of equilibrium that minimize entropy in order to attain homeostatic integrity. Homeostasis (inner-system equilibrium) is accomplished through the identification of sensory states by means of *ad hoc* receptors placed in a statistical boundary (*Markov Blanket*) providing a boundary between system-external and system-internal conditions.

This conceptual move is of great importance, as one of the motivations to compare biological and artificial systems is what I refer to as *increment bias*.[1] According to this

1 While artificial systems, such as large language models, may exhibit rapid advancements that seem to outpace human biological evolution, *increment bias* underscores a key difference between human and artificial

concept, artificial intelligence, specifically large language models, would exhibit rates of evolutionary advancement that surpass biological transitions regulated by environmental constraints. From this perspective, humans would be at a disadvantage compared to *in-silico* "cognizers" due to our constrained evolvability. However, the constitution of a semiotic Self is not the product of *reflection*, that is, the possibility of a computational architecture to "form representations of control flow that affect the control flow itself" (Barron et al., 2023, p. 5).

The chapter is divided into five main sections and a conclusion. I Section 1, I introduce the theoretical tenets of *Agentive Cognitive Construction Grammar*, a predictive theory of mind and language for the definition of the biophysical underpinnings of language. In Section 2, I offer and in-depth exploration of AgCCxG. In particular, I show how constructions possess a triadic structure combining form, function and agency that become active in specific events thanks to the action of constructional attachment patterns. In Section 4, I analyze the forms of essentialism encoded in LLM.

---

symbolic processing, namely that human cognition is deeply tied to our *semiotic Self*, where language is more than just a tool for communication—it's an adaptive, symbolic system that evolves within a context of lived experiences. Thus, central to an evolutionary increment bias is the need of stability, adaptability, and dependability of cognition, which artificial systems lack. This form of persistence allows human cognitive systems to remain stable and functional even within the slower evolutionary processes governed by environmental constraints. It is not merely the result of computational processes or control flow, as seen in artificial systems. Instead, human cognition and language endure because they are part of an intricate web of lived experiences, semiotic processes, and biological embodiment. Thus, the persistence of the human semiotic Self provides a form of cognitive stability that artificial systems, despite their rapid advancements, cannot replicate.

Among other things, I demonstrate that generated language sequences are for the most part generalizations from flattened content that lack embodied grounding, in contrast to natural language. In Section 5, I reveal the reasons why current usage-based approaches to language are at odds theoretically. One of the main reasons for this is the characterization of language as an intrinsically computational code for performance and statistical entrenchment of anecdotal constructionel combinations.

2. A predictive semiotic theory of mind

Agentive Cognitive Construction Grammar (AgCCxG) is a theory of mind and language that favors a strong embodied characterization of experience through a focus on concept formation and semiotic hybridization over time. An important consequence of this characterization is the definition of mind as a continuum of perceptual loops embedded in eventful states of consciousness. This is conditional upon precisely the existence of biological memory and entropy-reduction strategies. Another way of putting it is to say that uncertainty reduction, also known as free-energy minimization, also aims at facilitating the identification of the contours of the Self as an open biological system. The impetus for this claim is inspired by *Active Inference* (AIF), a process theory, adopted by different fields (e.g., computational neuroscience Friston, 2005, 2009, 2010; Clark, 2013; *Predictive Processing*, Clark. 2013; or mathematical frameworks such as the *Free-Energy Principle*, Friston 2010), as a means to provide a forma scaffolding to the biological imperative of free energy minimization. Among other things, AIF states that, in order to reduce *entropy* (that is, the breakdown of the contours of self-organizing systems), living organisms need to apply statistical predictions of the state of the world enabling them to cope with *surprise* (aka. free energy). In this chapter, AIF is hence considered as a framework for understanding how organisms actively gather information and take action in their environment to maintain a desired state or achieve a goal. This stresses the fundamental relationship between ground physical conditions supporting the organization and optimization of matter-energy exchanges within living organisms striving for priority and persistence in a specific Umwelt. This is expressed mathematically using Bayesian models that integrate probabilistic beliefs about the world and observations as a means to guide decision-making and action. The core AIF equations include:

1. **Bayes' rule**: P(h|d) = P(d|h) * P(h) / P(d) where P(h|d) is the posterior probability of a hypothesis h given data d, P(d|h) is the likelihood of the data given the hypothesis, P(h) is the prior probability of the hypothesis, and P(d) is the marginal probability of the data.

2. **The free energy principle**: F = - log P(d) + D_KL(Q(h|d) || P(h)) where F is the free energy, P(d) is the marginal likelihood of the data, Q(h|d) is the approximate posterior distribution over hypotheses given the data, and P(h|d) is the posterior distribution over hypotheses.

The Kullback-Leibler (KL) divergence between Q(h|d) and P(h) measures the distance between approximate and true posterior distributions.

3. **Action selection**: a* = argmax_a E_q(h|d)[log P(o|h,a)] where a* is the optimal action, q(h|d) is the approximate posterior over hypotheses given the data, P(o|h,a) is the likelihood of observations o given the hypothesis and action, and E_q(h|d)[ ] denotes the expected

value of the expression inside the brackets under the distribution q(h | d).

4. **Perception and inference**: q(h | o) = P(o | h) * exp(-F) / Z where q(h | o) is the posterior distribution over hypotheses given observations, P(o | h) is the likelihood of the observations given the hypothesis, F is the free energy, and Z is the normalization constant that ensures the distribution q(h | o) integrates to one over all possible hypotheses. This equation describes how organisms use their beliefs about the world and sensory observations to update their understanding of the environment and guide their actions.

I argue that this perceptual loop is first defined by specific modalities (interoception, exteroception, proprioception, or visceroception) in a *Markov blanket* (a term originally introduced by Pearl 1988, and later adapted by Friston 2010), and further organized by a modular structure of neuronal networks. It has to be said at this point that while Pearl's formulation of a Markov blanket is purely statistical, Friston's Markov blanket points to a true boundary between the internal and external states of a system, in a manner akin to a cell's membrane. For present purposes, however, a Markov blanket is not considered as a distinct boundary, but refers to the dynamic influence (energy exchange momentum) of an agent on an environment, whereby the system and the world can be distinguished from each other (the system does not possess a distinct boundary itself, though). This type of *open Markov blanket* can be formalized by using a Bayesian framework that incorporates both the agent's internal model of the environment and the actual sensory inputs they receive. The equations for the *open Markov blanket* can then be defined as follows:

1. **The agent's internal model of the environment:** $P(s_t \mid o_{1:t-1}, a_{1:t-1}, m)$ where s_t is the agent's internal state at time *t*, $o_{1:t-1}$ are the observed sensory inputs up to time t-1, $a_{1:t-1}$ are the actions taken by the agent up to time t-1, and m is the agent's model of the environment.

2. **The actual sensory inputs received by the agent:** $P(o_t \mid s_t, a_t, e)$ where o_t is the sensory input received by the agent at time t, s_t is the agent's internal state at time t, a_t is the action taken by the agent at time t, and e is the actual state of the environment.

3. **The agent's policy for selecting actions** based on its internal state and model of the environment: $P(a_t \mid s_t, m)$

4. **The dynamics of the environment**, which determine how the environment evolves in response to the agent's actions: $P(e_{t+1} \mid e_t, a_t)$

5. **The feedback loop** that updates the agent's internal state based on its actions and the environment's responses: $s_{t+1} = f(s_t, o_t, a_t, e_{t+1})$

In this formulation, the open Markov blanket refers to the dynamic influence of the agent on the environment, whereby the system and the world can be distinguished from each other. This dynamic influence is captured in the feedback loop that updates the agent's internal state based on their actions and environmental responses. In other words, *intelligent agents* make decisions about their action upon the world as a means to maintain equilibrium between the preservation of their model of the environment (Self-as-symbol) and the imperative to reduce uncertainty. This can be summarized in a working checklist:

1. *Intelligent agents* use language as a predictive tool for an adaptive action on the world.
2. Language bears traces of embodied content in the form of pre-semiotic cognitive routines.
3. Embodied experience is *direct* and *ecological* (that is, representational layers are not added to events in the brain).
4. Embodied cognition reflects a need to reconstruct events on the basis of bodily-acquired information.

3. Language as a predictive semiotic tool

According to this reading, language is a semiotic tool that reflects the dynamics of biological adjustment and self-preservation. The upshot is that there is a wide range of phenomena pertaining to language that involve learnability, actualization, and biological adjustment, and that can be translated functionally as the interplay between efficient and non-conflicting sign-combination and context-bound agentive decisions. This characterization requires us to view language as a sign system consisting of constructions encompassing three core elements: *form, meaning* and *intelligent agency* (see Figure 1). Under this view, constructions emerge as linguistic components for the conceptualization of an event defined by specific *constructional attachment patterns* (CAPs), that is, embodied constructional arrangements that organize both high-level (abstract syntactic patterns) and lower-level constructions such as words, morphemes, affixes, etc., in networks of interconstructional relations motivated an agent's need to offload intentional states onto specific contexts of use.

CAPs reveal different degrees of embodied experience with the physical world that provide concepts with unique tridimensional qualities (concepts are not mechanically retrieved from within a fixed symbolic system). As coefficients for semiotic relations among constructions, CAPs are driven by hypotheses for the reconstruction of events. This reconstruction process encompasses the integration of sensory and experiential aspects, as depicted by the formula below:

$$\text{CAP} = tc\ x\ er$$

where *tc* represents the weight of the connection between a *triadic construction tc* (see Figure 1.1) and an embodied representation *er*. These weights reflect the strength of the association between the linguistic and embodied components and can be entrenched through exposure and reinforcement. For example, the sentence "The sleeping snake in my belly woke up and started to slither" (expressing intransitive motion) can be analyzed as follows:

Argument Structure Construction	**Form**	**Intelligent Agency**
Intransitive Motion	(Subj)VOblique$_{\text{PATH}}$	IA_intransitive_motion **Formalism:** IntMot = IA + (Σw_k_im / TC_im)
	Meaning	
	X moves to Z	

Components:

- Argument structure construction: Describes

the syntactic structure and form of a specific construction.

- Form: The linguistic expression or pattern that represents the argument structure construction.

- Meaning: The semantic interpretation or function associated with the construction.

- Intelligent agency: Represents the influence or involvement of an intelligent agent in the construction.

- Equation: The equation that incorporates the intelligent agency factor and the associated triadic construction into the calculation of free energy.

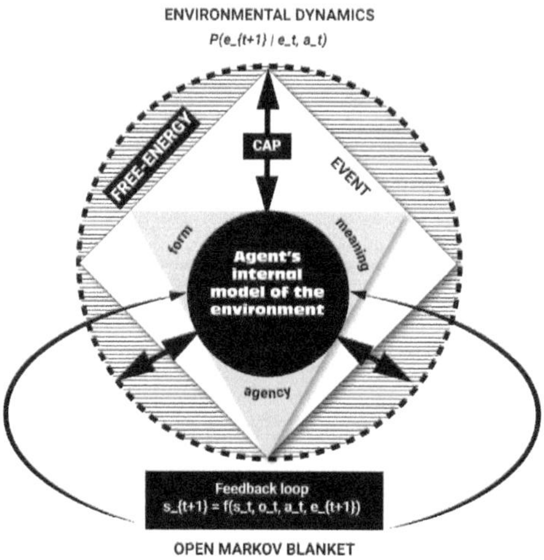

Figure 2.1. The model of triadic construction. The agent's internal model of the world provides the basis for the partition of events into discrete sequences of actions and conscious states that motivate the assembling of constructional elements though specific CAPs. CAP action aims at reducing free-energy by way of an estimation of the

state of the world measured through a feedback loop within an open Markov blanket.

Thus, instead of engaging in rough predictions on the basis of stored symbolic chunks, intelligent agents organize linguistic content by engaging in predictions about the outcome of specific usage within events, or chunks thereof. These predictions are motivated by world experience, rather than unmotivated, within-the-system linguistic permutations and generative self-organization. To assess this we need to explain what perceptual experience is supposed to be. In the first place, language is constructed on concepts, and concepts point to a need to recognize the substance of objects and entities in the world. However, essence identification is not construed as a simplistic assignment of value to a set of features on a continuum of prototypical characteristics possessed by entities and objects that could be described, for example, by way of single-layered propositions. Therefore, one main point of the present theory is to suggest that "unobservable essences" are not homogeneous, linearly arranged properties of identifiable natural kinds, but portions of experience retrievable across various layers of perceptual modeling in a bottom-up manner. The idea I am advancing here, then, is that language links these essences to observable surface characteristics through specific affordances which can be represented as a vector $E= [e_1,e_2,\ldots,e_k]$. On this construal, the essential properties of a natural kind, say, a tiger, is not simply a symbolic representation of the animal's intrinsic properties in a conceptualizer's mind, but a reflection of a network of potentialities inherent in the events where these essences are active.

Importantly, in defining the existence of an entity or object a multilayered set of propositions needs to be posited following this model: "If condition S becomes true upon the lower scale length ΔL, then condition S* will become true on the higher scale length ΔH" (Wilson, 2023, p. xiv-xv). This formulation underscores the intricate interplay between diverse perceptual scales and the multifaceted interpretation of properties as they relate to natural kinds[2] across these scales. From this perspective, a trait or property deemed essential for an agent's survival at one scale does not preclude its classification as an intrinsic property at another level. For instance, in discussing tigers, the absence of stripes may not inherently signify the core properties associated with tigerness. However, this absence may pose a disadvantage,

2 From this viewpoint, it becomes clear that brain ontologies related to language formation and processing cannot be assumed solely on established methodological and epistemic principles. This challenges the notion that identified brain regions provide definitive evidence of ontological reality (e.g., Fedorenko, Ivanova, & Regev, 2024). Likewise, claiming that a linguistic natural kind in the brain, loosely defined as "an ontologically meaningful grouping of brain areas" (Tuckute et al., 2024, p. 282), can be substantiated by the traditional encoding-decoding framework of symbolic string generation is untenable. Importantly, the symbolic output-oriented task design, influenced by the deficiency-fallibility bias, portrays human "performers" as "fallible but boundedly rational agents who can, in principle, overcome their ignorance if the task falls within cognitively permissible inference rules" (Solaki, 2022, p. 543; emphasis added). The preceding passage highlights a growing irony in modern algorithmic-driven epistemologies: while the original goal of the Turing test was to determine "whether someone could accurately identify if they were interacting with an AI or a human in a decontextualized setting" (Youssef et al., 2024, p. 6, citing Nov, 2023), the current focus has shifted to examining how well humans can create representations that enable efficient communication with machines using a limited set of predefined signs based on language model training. This algorithmic mindset has led to studies positing LLMs as "mind theorists" (Strachan et al., 2023), rationally-bounded agents (Yax et al., 2024), or human-like intuitive thinkers (Hagendorff et al., 2023), which points to an increased bias among many researchers toward LLM humanization and attribution of mental states. This has been mostly operated on the rationalistic premise that linguistic data alone can account for stable states of the world across natural language, which, as a result may induce a correct reconstruction of reference from raw text (e.g., Søgaard, 2024).

as it becomes a prerequisite at a related experiential layer, such as camouflage, and even more critical as an asset at the layer pertaining to survival through hunting. In this context, the fur property cannot be considered in isolation from its functional value in the natural environment. Similarly, the question of why propositions such as "Redness and squareness are intrinsic properties. Being next to a red object is extrinsic" (Vallentyne, 2014, p. 31) are rendered nonsensical by the multilayered nature of linguistic reconstruction invites a similar range of explanations.

Clearly, this reasoning runs out of wiggle room quickly, since the existence of stripes and orange fur (an evolutionary trait) is intrinsic to successful hunting (though irrelevant for coarse visual tiger identification), and may become an extrinsic trait for the Chital, whose sensory affordances[3] could, at some point, be deceived by the tiger's stripes, in which case this fully functional biological system becomes a potential prey. So, to conclude, it is not the case that "[w]hether something is red, or 3 kg, or round is a matter

3 According to Gibson (1969, 1977), embodied experience is regulated by environmental affordances, denoted by A, which serve as anchor points for action. An affordance is represented as a function $f: E \rightarrow A$, where E is the set of environmental features and A is the set of potential actions that an entity can exploit. Affordances are determined by the phylogenetic configuration of the entity, denoted by φ. Thus, an affordance can be defined as $A(\varphi, E) = f(E)$.
Objects in the environment possess affordances that can be utilized by entities based on perceptual cues such as weight (W), size (S), and form (F). These perceptually accessible properties influence the affordance function, resulting in the emergence of specific action profiles. Mathematically, this can be expressed as $A(W, S, F, E) = f(W, S, F, E)$. So we see that Gibson's emphasis on feature saliency highlights the centrality. This occurs thanks to semiotic mediators of embodied experience by which entities perceive and interact with their environment based on the potential actions available to them according to their phylogenetic configuration and perceptually accessible environmental features.

of how it itself is, regardless of anything else." (Denby, 2014, p. 91).

Moreover, at the heart of the tiger-chital affordance war, is the idea that local intrinsic properties are not merely parcels of experience brought to perceivers through modality-specific representations. We can summarize this assertion as follows:

1. Essence Identification and Perceptual Modeling:

- Model 1.1:
- Variables: $E=[e1,e2,...,ek]$ (Essence vector), $S=[s1,s2,...,sn]$ (Surface characteristics)
- Equation: $E=f(S)$
- Description: The essence vector E is a function of observable surface characteristics S, representing the retrieval of essences across diverse perceptual modalities.

2. Gradation and Transference of Properties:

- Model 2.1:
- Variables: ΔL (Lower scale length), ΔH (Higher scale length)
- Equation: $S*=g(S)$
- Description: Conditions at lower perceptual scales ΔL influence conditions at higher perceptual scales ΔH, indicating the transference of properties across perceptual layers.

3. Complexity and essential traits:

- Model 3.1:

- Variables: E_t (Essence vector for tigers), E_c (Essence vector for Chitals)

- Equation: $E_t=[stripes, fur\_color, \ldots]$, $E_c=[stripes, fur\_color, \ldots]$

- Description: Essential traits for tigers and Chitals contribute to the complexity of perceptual models, with traits like stripes and fur color influencing both intrinsic properties and functional significance within ecological contexts.

4. Contextual Relevance of Properties:

- Model 4.1:

- Variables: C (Contextual relevance)

- Equation: $C=h(S)$

- Description: Properties exhibit contextual relevance across perceptual scales, impacting both intrinsic and extrinsic attributions.

It is evident that this situation does not bode well for theories endorsing a linear, deterministic mapping of experience onto language, since the multifaceted nature of properties and their contextual relevance across various scales of perception does not allow for a one-dimensional assignation of content to concepts. As a result, it is not possible, to simply "ascribe properties to something" and then move on to state that considering a thing as being part of something larger is no longer a proposition about the thing itself but about something intrinsically different (e.g. Lewis, 1983). This is crucial to consider, especially when we acknowledge that the constructions comprising a language, along with the theories applied during their assembly in online discourse production, require more from the speaker than mere estimation of truth values (see Section 4). Clearly, the sum of linguistic routines, amendments, possible hesitations, creative expansion of semantic content, in short, the lived experience of surviving through a semiotic system tuned to the world, is not accessible to language models whose only imperative is to define statistically relevant patterns of symbolic combinatorics in a one-dimensional context of usage.

All of these considerations prompt us to ask: Is it feasible to disregard the intrinsic-extrinsic dichotomy while recognizing the nuanced intricacies of perception and the propositions that delineate the underlying processes? Describing this connection as one of "ontological dependence" (Marshall and Weatherson 2002), wherein the existence of one entity relies on the existence of another, presents its own set of challenges. Ontological dependence does not pertain to the structure of a system in relation to another system; instead, it represents a facet of the relationship between concepts and the perceiver/conceptualizer within a dynamic continuum of attentional states.

This leads us to suggest a definition of consciousness, denoted as C, as the understanding of an external reality (R) in conjunction with a constructed model of the Self (S) within a specific parcel of reality. Mathematically, this can be expressed as:

$$C=RUS$$

Essentially, this theoretical standpoint refutes the notion of the existence of internal representations. From this perspective, neither gorillas nor drosophilae possess a symbolic repertoire (SR) capable of being mapped onto real-world objects or entities (O). This can be represented as:

$$SR\not\leftrightarrow O$$

This is compatible with the idea that the strengthening of a sense of causal dependence, through the coupling of low/higher order perception and concept formation, is the hallmark of consciousness, rather than the ability to compute by means of a code enabling us to bring to ourselves a picture of the world. So, in the present theory, representation stems from the interplay of three types of concepts:

1. **Core concepts**: (Phylogenetically defined)

2. **Contingent concepts**: constructed upon the interaction of a system's affordance repertoire and the structure of the environment.

3. **Expanded concepts**: Updated patterns of agentic action upon an eco-niche that can be split into two components for non-cognizant open systems.

Natural language production, as discussed earlier, requires a broad range of experiences tied to our sense of self—something that artificial language models cannot replicate. This complex representation of the self in the world relies on the combination of our perceptions, how we conceptualize them, and the beliefs we project onto specific events in different contexts.

Take for example the verb "slither" in English.

This phrasal verb is typically associated with animate entities having a cylindrical body shape that constrains their movements in a certain way. In particular, snakes are said to *slither*, that is to move in either a rectilinear, lateral undulation (sliding contact without gripping), sidewinding, or concertina mode (periodic static gripping), in contrast to limbed animals (see Figure 1.2). The assignation of content during a "slithering event" thus entails an iconic reconstruction of body shape, skin texture, incline, surface roughness, as well as the rhythmic alternation of muscle flexion as a means to move forward. These core elements can be expanded metaphorically provided that shape and mode of locomotion do not create a conflict with the original embodied elements encoded in the verb. As a result, the particular selection of the proper verb in order to describe specific modes of locomotion depends on the assignation of affordances to an entity. Thus in the sentence, "The car silently *slithered away* **from** the curb", the sinuous manner of locomotion attributed to snakes (and also to snails) is projected onto a self-propelled object as a means to create a particular effect on the reader, whereby bodily experience, and not only statistically-driven construction combinatorics, is summoned. As shown in Figure 2, the genesis of concepts encoded in verbs is attributable to a battery of relations pertaining to the experience of agents with the physical world, and that may include well-entrenched essences associated with visual and motor modalities. These can be considered as core, essential elements that link iconic representation with body shape and a type of locomotion. However, as already mentioned, the embodied reconstruction of events can be distributed over several layers of experience where the description of specific phenomena require a nuanced treatment of what can be considered as an essential trait or an intrinsic property of a natural kind. For example, depending on the texture of a surface, or the structure of it, snakes can adopt either a concertina or an undulating mode of locomotion, which reveals that "to slither" is an essentialist term that cannot capture all of the possible facts about that particular event of locomotion. As a result, to "silently slither away" is not a property of cars but a manner of movement attributed by an agent to the vehicle in that particular event, which reveals that the distinction between what is essential and intrinsic is also the product of the intention of the agent conceptualizer.

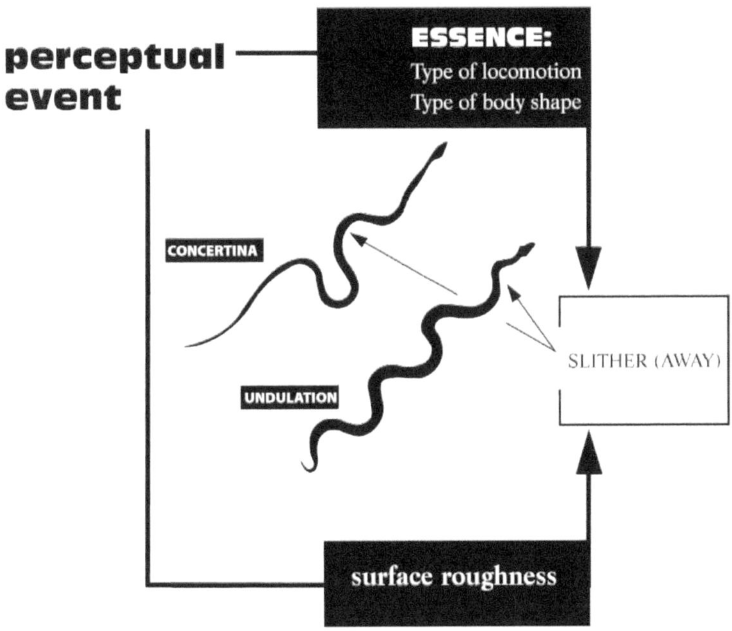

Figure 2.2. A succinct model of iconic essence mapping in language. As shown, both surface structure and body shape are encoded in the verb "slither [away]" as a means to describe a specific type of movement associated with cylindrical body shapes.

4. Large language models' statistical essentialism

The spectacular advancement in *natural language processing* (NLP) in the field of artificial intelligence has led to the consolidation of language models like the GPT (Generative Pre-trained Transformer) series. This research has also impacted our views of language and intelligence. In particular, formal linguistic models of natural language have been challenged by the capacity of transformers to generate language without recourse to syntactic trees, innately given rules, or specific form-function pairings (constructions), as postulated by usage-based approaches (Goldberg 2006). An important feature of these models is that they undergo extensive pre-training on massive corpora of text data, which results in remarkable performance across a wide spectrum of NLP tasks, ranging from language translation and text summarization to question-answering. ChatGPT stands out for its exceptional performance and versatility, showcasing promise across various domains including education (Frieder et al., 2023; Kim & Su, 2024) and healthcare (Imperial et al., 2024). Its multifaceted utility underscores the transformative potential of advanced NLP technologies in addressing real-world challenges and enhancing human-computer interactions.

In the case of ChatGPT, a language model that generates context-specific language according to a user's prompts, the process of generating sentences that reflect real-world physics, as well as the semantic nuance introduced by a human speaker, is divided into eight steps whereby language is transformed into symbols that fit a pre-existing statistical model of semantic distribution and syntactic arrangement. In the case of the phrasal verb "to slither away", the prompt sentence "the car silently slithered away from the curb" is first "tokenized", that is, the sentence is chunked into individual words: "The", "car", "silently", "slithered", "away", "from", "the", "curb". Then, each token is mapped to an *embedding vector* in a high-dimensional space. For example, the embedding for "car" might be represented as *car*$_e$*car*. Input tokens are then processed in *ChatGPT's Transformer architecture* wherein the input tokens go through multiple layers of self-attention and feed-forward neural networks. An attention mechanism helps the model to focus on relevant parts of the input sequence. For instance, when generating the word "slithered", the model may pay more attention to the preceding words "car silently". In the next instance, "training", ChatGPT uses a large corpus of text data "to learn" and ultimately predict the next token given the preceding context. For example, it learns that "slithered" often follows descriptions of movement like "silently" (the reconstruction of the event is, hence, purely statistical).

In this sense, when generating text, ChatGPT predicts the next token based on its learned probabilities. For instance, given the context "The car silently", the model might predict "slithered" with high probability. After generating the sequence of tokens, *detokenization* reconstructs the original sentence from the tokenized form. For example, the model merges sub-words to form complete words and handles punctuation. In this case, detokenization would convert the tokens back into the original sentence: "The car silently slithered away from the curb". Finally, the generated text may undergo further post-processing such as formatting adjustments (see Figure 1.3).

(1)

a) The snake silently ***slithered away into*** the dense undergrowth.

b) After being startled, the lizard quickly ***slithered away under*** the rock.

c) Sensing danger, the eel gracefully ***slithered away from*** the approaching predator.

d) As the light faded, shadows seemed to ***slither away into*** the darkness.
e) Noticing the suspicious movement, the spy discreetly ***slithered away from*** the crowded room.
f) The thief attempted to grab the treasure, but it seemed to ***slither away from*** his grasp.
g) With a flick of its tail, the catfish ***slithered away from*** the fishing line, escaping into the depths of the pond.
h) Despite its size, the slug could effortlessly ***slither away from*** potential threats.
i) The fog slowly began to dissipate, its tendrils seeming to ***slither away with*** the rising sun.
j) Startled by the sudden noise, the insects ***slithered away from*** the fallen branch, seeking shelter in the nearby foliage.

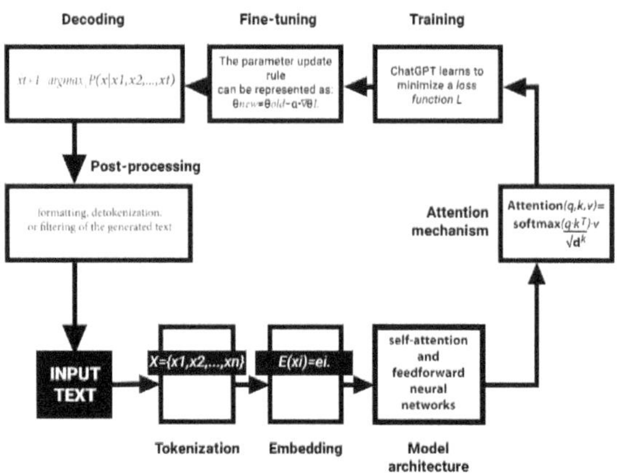

Figure 2.3. The cycle of language generation of ChatGPT. The process of language generation follows specific steps. Tokenization involves splitting the input text X into individual tokens, represented as X={x1,x2,...,xn}. Each token xi is then transformed into a high-dimensional vector representation, ei, through embedding: E(xi)=ei. ChatGPT utilizes a Transformer-based architecture, comprising layers of self-attention and feed-

forward neural networks, with each layer's output computed as $Layeri(input)=LayerNorm(MultiHeadAttention(input)+input)$. Within this architecture, attention mechanisms enable the model to focus on different parts of the input sequence, computed as $Attention(q,k,v)=softmax(q\cdot kT/\sqrt{}dk)\cdot v$. During training, ChatGPT learns from text data, minimizing *cross-entropy loss*: $L(Y,Y\char`^)=-\Sigma iyilog(y\char`^i)$. Optionally, fine-tuning on domain-specific data may occur, updating parameters via $\theta new=\theta old-\alpha\cdot\nabla\theta L$. Decoding involves selecting the next token $x\char`^t+1$ based on its conditional probability given the preceding context: $x\char`^t+1=argmaxxP(x|x1,x2,...,xt)$. Finally, generated text may undergo post-processing steps such as detokenization, filtering, or formatting.

As we see in example 1a, ChatGPT can aptly predict verb usage from a simple prompt sentence. However, the model's responses are limited to a fixed syntactic pattern that, though grammatically correct, does not necessarily capture event structure properly. Since the prompt was "Write ten sentences using this sentence as a model",[4] ChatGPT

4 The analysis of language generation in Large Language Models reveals fundamental limitations in their ability to mimic human communication. While these models demonstrate an impressive command of linguistic fluency, they rely on statistical prediction rather than the grounded understanding of world knowledge that human speakers mobilize for meaningful interaction. This disjunction between linguistic *form and meaning*, where form becomes ***an allegory for understanding,*** offers a critical lens for examining the outputs of LLMs. For example, as stated by Habermas (1989), by employing insights from *Adorno's critical theory* and Brecht's dialectical approach, LLMs' generated language can be construed as a semiotic system operating in the realm of *allegory*—a disembodied system of symbols that points to meaning but lacks genuine comprehension:

> "The unity of sign and meaning, particularity and generality in the symbol, contrasts with the concept of allegory, which is based on the separation of sign and meaning. Given that modern society is based on the principle of separation, it is tempting to view allegory as the modern form par excellence

recognized the pattern but quickly run out of suitable content to fit the particular event configuration suggested by the prompt sentence (*model collapse*, that is the generation of repetitive, one-sided, or incoherent output as a result of disproportionate data points in the training data, see *Federal Bureau of Information Security*, 2023, p. 10).[5] This points to

and the symbol as merely the remainder of a premodern attitude. The persuasiveness of this categorization, however, is based on an abstract classification of thinking. If, on the other hand, one engages with the movement of form, one cannot remain with the simple juxtaposition of (modern) allegory and (premodern) symbol but must investigate their relationship as a whole." (Habermas, 1989, p. 90).

Seen through this lens, one of the most striking characteristics of LLM outputs is their allegorical nature, particularly in their handling of verbs of motion. As we have seen, LLMs produce linguistic structures that seem meaningful but remain unmoored from embodied knowledge. Therefore, command definition by a user is defined by the profound disconnection between form and content—this same disconnect is evident in LLM-generated text. So we see that allegory becomes the process by which LLMs create representations of the world that are statistically grounded but conceptually "unanchored". For instance, when an LLM generates sentences like "the snake slithered away," it is relying on statistical patterns rather than an understanding of the biological and physical properties of snakes. The model can replicate such patterns because they are abundant in its training data, but this replication is merely an allegorical gesture. The LLM does not access the embodied knowledge that informs human language, nor does it comprehend the real-world implications of the word "slither" beyond its linguistic context. As recent research suggests (e.g., Weissweiler et al., 2023), this reliance on disembodied statistical predictions points to a fundamental limitation of LLMs, which operate through form but without grounding in true world knowledge.

5 Model collapse is one key aspect of LLM's limited language generation loop during semiosis. In particular, we can see in this limitation of data processing an intrinsic lack of teleological quality, that is, a failing sense of movement toward the "ultimate", the *telos*, considered here as the accomplished reading of an event as grasped through the senses and, ultimately construed as experience. Since LLMs cannot one revealing characteristic of LLM output, namely its allegorical nature. As noted by Adorno, allegory is marked by a disconnection between form and meaning. In this sense, LLMs generate linguistic structures that seem meaningful at the surface level but lack the cognitive grounding necessary for true comprehension. This disconnect becomes particularly evident in the model's handling of verbs of motion, such as "slither away," where physicality is implied, but the underlying semantic understanding of what it means to slither is absent. Thus, sentences (1a,b,c), focus on the intrinsic properties of a type of movement associated with specific body shape. Samples (1d,e,f,i) introduce metaphorical expansion that, nonetheless, can be hardly licensed by embodied world knowledge, that is, they defeat the purpose of using the verb as a means to describe the behavior of a specific animated object or entity in the reported event (the transformer introduced the hedging verb "seem" in cases where the usage of the verb "slither away" was not warranted by the prompt sentence).

In other words, the semantics of the verb in the generated sentence is not expanded but constrained by the prompt in the model's representational space. Interestingly, in the case of (1i), the creative use of "slither away" results from the introduction of the word "tendrils", which creates a layer of animacy for an atmospheric phenomenon (the fog) that aligns with the form+way of locomotion real-world bias.

---

be both Interpretants within and outside semiosis, it is unlikely that their representations can be further expanded to a continuous process of meaning reconstruction in events defined by world knowledge and embedded experience, In LLM-training, this basically means that a LLM being trained by data provided by another LLM fails to recognize the very training task, as no semiosis can be identified from data retrieved from the introduction of improbable sequences generated by the models themselves (see Shumailov et al., 2024).

Likewise, sample (1g) properly captures the event in which "active gripping" is counteracted by lateral undulation as an element encoded in the slithering action performed by the fish; however, sample (1h) fails to create a plausible event, since the adverb "effortlessly" cannot be freely associated with an escaping action performed by a slug (remember that ChatGPT predicts verb usage by paying attention to –*ly* adverbs placed before the target verb); finally, in (1j), the model creates a conflict between the event depicted and human world-knowledge by attributing a "*slithering action*" to an unsuitable agent (insects actually "crawl", which presupposes a "limbed" type of locomotion).

So it is evident that the disembodied, unconstrained statistical prediction of pre-trained language generators cannot emulate the human need to mobilize world knowledge as a means to provide accurate, context-oriented representations of the world. Furthermore, current architectures and learning setups are still poorly equipped to creatively construct sentences that reflect the actual intention of an agent engaging in a type of communication not activated by a prompt. As we have seen in the sentences provided by the Language Model, language generation and semantic understanding provide graded semiotic access to meaning.

5. Construction grammars and pre-trained transformers

Despite challenges, research persists in applying linguistic methodologies to analyze language generation by *Large Language Models* (LLMs). Notably, usage-based approaches such as construction grammars (Goldberg, 2006; Cappelle, 2024) have garnered attention in this context with efforts being made to bridge LLM development with linguistic inquiry (Beuls & Van Eecke, 2024; Weissweiler et al., 2023a,c). However, incorporating construction grammar predictions into LLM training remains problematic. One main reason is that, since its inception by Fillmore (1988), construction grammars have struggled to establish themselves as a coherent and ontologically robust discipline, particularly in contesting the Chomskyan linguistic paradigm. However, despite offering intriguing insights into linguistic nuance (Brisard 2023), construction grammars have generally fallen short in providing a comprehensive understanding of language and language usage. As a result, linguistic representation is often reduced to questionable simplifications, due to an over reliance on statistical learning methods (e.g., Goldberg 1995, 2006, 2019). The dearth of substantive insights into the cognitive essence of constructions (see Silvennoinen, 2023, for a discussion) has relegated construction grammars to the realm of speculative constructs (Samuel, 2020). This is particularly evident in areas where the application of construction grammar's conceptual tools fails to substantiate robust assertions. For instance, endeavors to reduce modal meaning to formulas identified in corpora, aimed at simplifying explanatory complexity, have proven inadequate (Depraetere et al. 2023). Moreover, the lack of cohesive theoretical underpinnings within the constructionist community has given rise to the formulation of arbitrary "principles" (e.g. Leclercq & Morin, 2023).

As discussed in section 3, the cognitive dimension of language arises from recognizing that linguistic decisions are driven by the imperative of biological open systems to uphold equilibrium between Self-as-symbol and the environmental dynamics. This implies that the entire framework of propositions defining human-world interactions should be organized into a hierarchy of potential models of the world, each associated with specific linguistic content. In this context, a proposition featuring a natural kind, such as a tiger, should refrain from treating the essential nature of tigerness as an intrinsic property, without further considerations regarding the potential relationship of this concept with other natural kinds that may be part of the proposition's content. Consequently, the hunting behavior of a tiger (encompassing stalking, ambush hunting, or swimming), irrespective of the feline's physical characteristics such as body configuration, fur color, and predatory instincts, should also be described in terms of the bodily affordances presented by its prey that define its being in the same eco-niche. Therefore, the usage-based claim that there is absolutely no trace of evolution or biological *persistence* in the construction of language is untenable.[6] As we have seen, the eons of experience of our species on the planet do not become accessible to us simply through statistical bias. Although languages are not innate *per*

6 The term "persistence" refers to the stability and dependability of our cognitive functionality, which is crucial for survival and adaptation. In this view, language is one of the key cognitive tools that support this stability. On this reading, human cognition, and especially language, functions in a way that ensures the continuous ability to interpret, interact with, and respond to our environment effectively. Language, in this sense, is not just a means of communication, but a core cognitive mechanism that allows for predictive processing—the ability to anticipate and understand events, behaviors, and contexts. This capacity to predict and interpret the world in symbolic terms is part of what sustains our cognitive functions, ensuring their "persistence" over time. Hence, unlike artificial models, which lack true experiential grounding, human cognition, through language, maintains this continuity by constantly adapting to new experiences, perceptions, and conceptual frameworks. Otherwise stated, "persistence" refers to the cognitive system's robustness and adaptability, enabling language to continuously function as a tool that both reflects and shapes human thought, aiding survival through the minimization of uncertainty in a complex world.

se, the mechanisms for concept formation and manipulation are innately given and further refined by experience and agency. On this reading, language evolution does not occur within the heads of individuals through semiotic mutations within the system, surfacing in anecdotal exchanges without other motivation than "sneezing foam off our cappuccinos," but as a manifestation of larger representational events that summon the whole host of experiences that define us humans. The key element here is phylogeny, which can hardly be considered as a rhetorical move for the positioning of an idealized speaker.

Indeed, although it is evident that both construction grammars and AI research depart from a reduced view of human experience, pre-trained transformers do not need constructions to generate language (see Weissweiler et al. 2023b). Notwithstanding, it cannot go unnoticed that both construction grammars and LM research confine world knowledge to the positing of propositions through which real-world phenomena are tokenized and resemiotized according to predefined rules of communication, which disregards conceptual groundedness (the process of ecological concept formation through language), embedding (the embodied, context-driven reconstruction of an event in context), and situatedness (eventfulness). As we have seen, this offers an inaccurate reading of language as a tool for the positioning of speaker intention. As a result, even when construction grammarians acknowledge the role of agency in the emergence of constructions (Goldberg, 2019), it is only to reinforce the idea that it is the linguistic system, rather than its purpose which defines the roles of speakers during communication.

6. Conclusion

In times where linguistics is increasingly compelled to embrace the algorithmic view of language, through the reduction of human representation to data structures for the acquisition of theoretical foothold (as seen in the development of computational construction grammar, Beuls and Van Eecke, 2024), it is crucial for researchers in linguistics to envision more accurate models of human representation through language. In this sense, human linguistic data cannot be reduced to instances of LLM performance, such as perplexity values. Another point worth considering here is whether training a transformer to recognize specific constructions (e.g., caused motion constructions, Weissweiler et al. 2023c) may at some point help buttress the theoretical plausibility of the notion of form-function pairings in general. Since argument structure constructions are abstract syntactic patterns that reflect general trends in the human construction of world knowledge, it is of little relevance for linguists to explore how "peripheral" constructional content (such as caused motion constructions) could at some point confirm the existence of constructions as ontological objects contributing to the optimization of LLM performance (e.g., Weissweiler et al., 2024).

As we have seen, a focus on the formal properties a language has in virtue of what it is in isolation has led to confusion regarding the role of natural language as a model for the development of pre-trained transformers. Certainly, language has a number of properties that are not purely "linguistic", but both biological and physical. Moreover, performance and computational efficiency are not intrinsic properties of language. If they were, computers and humans would be indistinguishable, as their sole function would be to combine symbols to externalize logical propositions about the state of the world. This stresses the underlying interconnectedness between different perceptual scales and the potential transference of properties across these scales. Among other things, it shows that a trait deemed essential for an agent's survival at one scale does not negate its status as an intrinsic property defining its way of being at a higher level, where the value of condition S^* is influenced by a modified set of relations. So, if intrinsic properties were to be had in language production/generation, these would be dependent on the structure of the experiential parcel defined by an event and not a particular instance of formally interpreted linguistic performance. As I have indicated previously, the fact that natural language is a tool to (re) construct experience for purposes other than language itself shows that linguistic intrinsicality is defined by agentive imperatives at every step of the process, and that what is intrinsic about concept formation is not what we make of natural kinds or the objects in the world, but our ability to distinguish beliefs from facts. In contrast, pre-trained transformers are not conscious agents, but performative ones, the nature of their responses being restricted by their capacity to identify efficiently the content of a prompt. Thus, while transformers exhibit superior computational capacity, humans excel at purposeful biological adjustment and optimization.

One of the lessons to be learned from the discussion introduced in this chapter is that LLMs' linguistic capabilities do not say anything substantial about the way we think. Indeed, in terms of chatbot-human interaction, we should refrain from suggesting that transformers are in possession of *concepts about concepts*, especially regarding human metalinguistic knowledge. As already noted, transformers are trained and supervised by specific human agents according to specific cultural, racial, and corporate criteria that condition what these "stochastic parrots" can or cannot say. Moreover, the depletion of high-quality linguistic content to train LLMs by 2024 (Maslej et al., 2024, p. 52) announces an increment in data reuse in the form of synthetic data (LLM-generated text to train another LLM), which points to increased levels of AI language essentialism. Therefore, taking bot essentialism is often the result of *automation bias*, that is, excessive credibility on the part of users due to the bot's correctness and convincing text output (see *Federal Bureau of Information Security* 2023, p. 10), Consequently, automation bias can be defined as a cognitive process by which users of chatbots ascribe sentience and epistemic value to the responses of a language model in the belief that these are actual agents possessing awareness of nuance and a sense of appropriateness during a communicative event:

> On a non-academic level, ChatGPT may *teach users various (meta)linguistic lessons*, ranging from basic language use (in terms of semantics and syntax, to start with) to more subtle and nuanced aspects, involving for instance politeness" (Dynel 2023, p. 122; my emphasis).

In the above passage, the author attributes *essential properties* to the Chatbot by suggesting that its adjustments through training and enhanced processing capabilities create a form of ontological dependence (inherent in the linguistic interactions between human agents) between the user's needs and the prompt-driven event governing the generation of "polite" language. As previously discussed, propositions that lack gradation in describing events with agential participants, oversimplify the hierarchy of those events through binary predictions of outcomes and relations (e.g. Cantlon & Piantadosi, 2024). As demonstrated in example (1), the author makes a prediction that, nevertheless, imbues the machine with psychological content to support the assertion that chatbots have a sense of language in their interaction with humans. However, *politeness* is not an intrinsic property of the chatbot activated by its interaction with a human, but a feature introduced during transformer training by a human individual with specific cultural biases and interests (thereby resulting in statistical bias, a way of behaving). An essential aspect to consider regarding human perception of AI-generated language is the oversight of the layers of abstraction inherent in computational artifacts. These layers encompass, among others, the designer's intentions for the solution formulated to address a given problem. It is crucial to recognize that displaying politeness in behavior (the "*how*") does not inherently equate to possessing an intrinsic quality of being polite (the "*what*"). Consequently, suggesting a naturalistic alignment of intentions involving meaning negotiation and pragmatic adjustment becomes implausible when only one participant, the human agent, possesses awareness of their conscious states enabling them to plan beyond the immediate chatting event, which is determined by the prompt imposed by the human.

A misinterpretation of this fact has led to hasty conclusions regarding the metalinguistic and metapragmatic affordances of chatbots, which can only be interpreted as default responses vis-a-vis users' misuse of the tool, often for entertainment purposes. In fact, the statement "as an AI language model", used as a means to introduce a disclaimer regarding the limits of its training or its world knowledge (Dynel, 2023), can be hardly interpreted as an indication of selfhood awareness for the production of context appropriate discourse during its interaction with a user. Regarding the nature of user evaluation of the chatbot performance, it is also a stretch to assert that most users are equipped with the appropriate metapragmatic, metalinguistic, or metadiscursive tools to assess the bot's competence beyond their own disruptive behavior in online interactions. This is of utmost importance when we acknowledge that the patterns of interaction among human agents on social media and chats are marked by an undue reaffirmation of their own beliefs and prejudices through destabilizing and tendentious language usage. Since these agents are always seeking to vet gratuitously other people's capacities, knowledge base, and reactions, it is not an insight to assert that this proclivity could be extended to human-chatbot interaction as well. This "dummy effect" (my coinage), that is, the tendency to engage in abusive language use during interactions with chatbots, also brings to the fore the various ways linguistic (mis)use can be exerted in the near future on robots. A point worth considering here pertains to the appropriate distinction between terms and their application to describe particular interactions mediated by language.

Since transformers exhibit different representational structures which can be aptly dubbed as a distinct computational semiosis possessing a surface human-like structure, but lacking experiential layers anchored in the physical world, the emergence of new types of conceptualizers that may eventually define new forms of interaction among biological and artificial agents possessing a symbolic contour and goals requires us to redefine our roles in an increasingly complex environment where our success can no longer be taken for granted. The semiotic Self entails the reemergence of the human, a being that, in order to claim their humanity, does not need to "exclude his (sic) own strangeness or animality" (Zalloua, 2021, p. 4).

References

Barron, A. B., Halina, M., & Klein, C. (2023). Transitions in cognitive evolution. *Proceedings of the Royal Society B, 290,* 20230671.

Beuls, K., & van Eecke, P. (2025). Construction grammar and artificial intelligence. *Preprint.* To appear in M. Fried & K. Nikiforidou (Eds.), *The Cambridge Handbook of Construction Grammar.* https://doi.org/10.48550/arXiv.2309.00135

Birch, J., Ginsburg, S., & Jablonka, E. (2020). Unlimited associative learning and the origins of consciousness: A primer and some predictions. *Biology & Philosophy, 35*(56).

Braidotti, R. (2013). *The posthuman.* Cambridge, UK: Polity Press.

Brisard, F. (2023). Spectacle and sensationalism in Construction Grammar. *Constructions, 15*(1). https://doi.org/10.24338/cons-536

Cantlon, J. F., & Piantadosi, S. T. (2024). Uniquely human intelligence arose from expanded information capacity. *Nature Reviews Psychology.* https://doi.org/10.1038/s44159-024-00283-3

Cappelle, B. (2024). *Can construction grammar be proven wrong?* Cambridge: Cambridge University Press.

Clark, A. (2013). Whatever next? Predictive brains, situated agents, and the future of cognitive science. *Behavioral and Brain Sciences, 36*(3), 181–204.

Clarke, B. (2008). *Posthuman metamorphosis: Narrative and systems.* New York: Fordham University Press.

Collins, A., & Quillian, M. R. (1969). Retrieval time from semantic memory. *Journal of Verbal Learning and Verbal Behavior,* 8(2), 240–247.

Coole, D., & Frost, S. (Eds.). (2010). Introducing new materialisms. In *New materialism: Ontology, agency and politics* (pp. 1–43). Durham: Duke University Press.

Demuro, E., & Gurney, L. (2024). Artificial intelligence and the ethnographic encounter: Transhuman language ontologies, or what it means "to write like a human, think like a machine." *Language & Communication,* 96, 1–12.

Denby, D. (2014). Essence and intrinsicality. In R. Francescotti (Ed.), *Companion to intrinsic properties* (pp. 87–110). Berlin / Boston: Walter De Gruyter.

Depraetere, I., Cappelle, B., & Hilpert, M. (2023). Introduction. In I. Depraetere, B. Cappelle, M. Hilpert, L. De Cuypere, M. Dehouck, P. Denis, S. Flach, N. Grabar, C. Grandin, T. Hamon, C. Hufeld, B. Leclercq, & H.-J. Schmid (Eds.), *Models of modals: From pragmatics and corpus linguistics to machine learning* (pp. 1–13). Berlin / Boston: De Gruyter Mouton.

Dynel, M. (2023). Lessons in linguistics with ChatGPT: Metapragmatics, metacommunication, metadiscourse, and metalanguage in human-AI interactions. *Language & Communication, 93,* 107–124.

Federal Bureau of Information Security. (2023). Generative AI models: Opportunities and risks for industry and authorities. Retrieved from https://www.bsi.bund.de

Fedorenko, E., Ivanova, A., & Regev, T. (2024). The language network as a natural kind within the broader landscape of the human brain. *Nature Reviews Neuroscience, 5*, 289–312.

Fillmore, C. J., & O'Connor, M. C. (1988). Regularity and idiomaticity in grammatical constructions: The case of *let alone*. *Language, 64*(3), 501–538.

Frieder, S., Pinchetti, L., Chevalier, A., Griffiths, R.-R., Salvatori, T., Lukasiewicz, T., Petersen, P. C., & Berner, J. (2023). Mathematical capabilities of ChatGPT. *arXiv Preprint*. https://doi.org/arXiv:2301.13867

Gärdenfors, P. (2007). *How Homo became sapiens: On the evolution of thinking.* Oxford: Oxford University Press.

Goldberg, A. E. (1995). *Constructions: A construction grammar approach to argument structure.* Chicago: University of Chicago Press.

Goldberg, A. E. (2006). *Constructions at work: The nature of generalization in language.* Oxford: Oxford University Press.

Gärdenfors, P. (2023). The evolution of meaning spaces: How cognition drives language. *Cognitive Science, 47*(4), e13270. https://doi.org/10.1111/cogs.13270

Ginsburg, S., & Jablonka, E. (2019). *The evolution of the*

sensitive soul: Learning and the origins of consciousness. Cambridge, MA: MIT Press.

Goldstein, E. B. (2024). *Cognitive psychology: Connecting mind, research, and everyday experience* (6th ed.). Boston: Cengage Learning.

Harnad, S. (1990). The symbol grounding problem. *Physica D, 42*(1–3), 335–346.

Hilpert, M. (2019). *Construction grammar and its application to English.* Edinburgh: Edinburgh University Press.

Hodges, B. H. (2007). Good prospects: Ecological and social perspectives on conforming, creating, and caring in conversation. *Language Sciences, 29*(5), 584–604.

Hoffmeyer, J. (2008). *Biosemiotics: An examination into the signs of life and the life of signs.* Scranton: University of Scranton Press.

Hollis, G. (2024). Learning representations: From predictive processing to construction grammar. *Language, Cognition, and Neuroscience, 39*(8), 891–915.

Holyoak, K. J., & Morrison, R. G. (Eds.). (2012). *The Oxford handbook of thinking and reasoning.* Oxford: Oxford University Press.

Hume, D. (1739). *A treatise of human nature.* London: John Noon.

Jackendoff, R. (2002). *Foundations of language: Brain, meaning, grammar, evolution.* Oxford: Oxford University Press.

Jäkel, O., & Ziem, A. (2023). Social cognition and grammar: A constructionist perspective. *Journal of Cognitive*

Linguistics, 34(3), 355–380.

Langacker, R. W. (1987). *Foundations of cognitive grammar: Theoretical prerequisites* (Vol. 1). Stanford, CA: Stanford University Press.

Langacker, R. W. (2008). *Cognitive grammar: A basic introduction*. Oxford: Oxford University Press.

Levinson, S. C. (2022). *Interactivity in human communication*. Cambridge: Cambridge University Press.

Levinson, S. C., & Wilkins, D. P. (Eds.). (2006). *The roots of human sociality: Culture, cognition, and interaction*. Oxford: Berg.

Linell, P. (2009). *Rethinking language, mind, and world dialogically*. Charlotte, NC: Information Age Publishing.

Lupyan, G., & Bergen, B. (2016). How language programs the mind. *Topics in Cognitive Science, 8*(2), 408–424.

Maturana, H. R., & Varela, F. J. (1980). *Autopoiesis and cognition: The realization of the living*. Dordrecht: D. Reidel Publishing Company.

Merleau-Ponty, M. (1945). *Phénoménologie de la perception*. Paris: Gallimard.

Peirce, C. S. (1931–1958). *Collected papers of Charles Sanders Peirce* (Vols. 1–8). Edited by C. Hartshorne, P. Weiss, & A. W. Burks. Cambridge, MA: Harvard University Press.

Rorty, R. (1989). *Contingency, irony, and solidarity*. Cambridge: Cambridge University Press.

Tomasello, M. (2019). *Becoming human: A theory of ontogeny.* Cambridge, MA: Harvard University Press.

Tuckute, G., Kanwisher, N., & Fedorenko, E. (2024). Language in brains, minds and machines. *Annual Review of Neuroscience, 47.* https://doi.org/10.1146/annurev-neuro-120623-101142

Vallentyne, P. (2014). Intrinsic properties defined. In R. M. Francescotti (Ed.), *Companion to intrinsic properties* (pp. 31–40). Berlin/Boston: Walter de Gruyter.

Varela, F. J., Thompson, E., & Rosch, E. (1991). *The embodied mind: Cognitive science and human experience.* Cambridge, MA: MIT Press.

Watson, R. (2023). Agency, goal-directed behavior, and part-whole relationships in biological systems. *Biological Theory, 19,* 22–36.

Weissweiler, L., He, T., Otani, N., Mortensen, D. R., Levin, L., & Schütze, H. (2023a). Construction grammar provides unique insight into neural language models. In *Proceedings of the First International Workshop on Construction Grammars and NLP (CxGs+NLP, GURT/SyntaxFest 2023)* (pp. 85–95). Washington, D.C.: Association for Computational Linguistics. Retrieved from https://aclanthology.org/2023.cxgsnlp-1.10/

Weissweiler, L., Hofmann, V., Kantharuban, A., Cai, A., Dutt, R., Hengle, A., Kabra, A., Kulkarni, A., Vijayakumar, A., Yu, H., Schütze, H., Oflazer, K., & Mortensen, D. R. (2023b). Counting the bugs in ChatGPT's wugs: A multilingual investigation into the morphological capabilities of a large language model. In *Proceedings of the 2023 Conference on Empirical Methods in Natural*

Language Processing (pp. 6508–6524). Singapore: Association for Computational Linguistics.

Weissweiler, L., Köksal, A., & Schütze, H. (2023c). Hybrid human-LLM corpus construction and LLM evaluation for rare linguistic phenomena. *arXiv:2403.06965*.

Wilson, M. (2023). *Imitation of rigor: An alternative history of analytic philosophy.* Oxford: Oxford University Press.

Wittgenstein, L. (1953). *Philosophical investigations.* Oxford: Blackwell.

Yax, N., Anlló, H., & Palminteri, S. (2024). Studying and improving reasoning in humans and machines. *Communications Psychology, 2*(51). https://doi.org/10.1038/s44271-024-00091-8

Youssef, A., Stein, S., Clapp, J., & Magnus, D. (2023). The importance of understanding language in large language models. *The American Journal of Bioethics, 23*(10), 6–7.

Zalloua, Z. (2021). *Being posthuman: Ontologies of the future.* London/New York: Bloomsbury Academic.

CHAPTER 3

Dehumanizing the human, humanizing the machine: Organic consciousness as a hallmark of the persistence of the Human against the backdrop of artificial intelligence

Abstract

The rise of *generative artificial intelligence* (GenAI), especially *large language models* (LLMs), has fueled debates on whether these technologies genuinely emulate human intelligence. Some view LLM architectures as capturing human language mechanisms, while others, from a posthumanist and a tranhumanist perspective, herald the obsolescence of humans as the sole conceptualizers of life and nature. This paper challenges, from a *practical philosophy of science* perspective, such views by arguing that the reasoning behind equating GenAI with human intelligence, or proclaiming the "demise of *the human*," is flawed due to conceptual conflation and reductive definitions of humans as performance-driven semiotic systems deprived of agency. In opposing theories that reduce consciousness to computation or information integration, the present proposal posits that consciousness arises from the holistic integration of perception, conceptualization, intentional agency, and self-modeling within biological systems. Grounded in a model of *Extended Humanism*, I propose that human consciousness and agency are tied to biological embodiment and the agents' experiential interactions with their environment. This underscores the distinction between pre-trained transformers as "posthuman agents" and humans as purposeful biological agents, which emphasizes the human capacity for biological adjustment and optimization. Consequently, proclamations about human obsolescence as conceptualizers are unfounded, as they fail to appreciate intrinsic consciousness-agency-embodiment connections. One of the main conclusions is that the capacity to integrate information does not amount to phenomenal consciousness as argued, for example, by *Information Integration Theory* (ITT).

Keywords: Consciousness, concept formation, extended humanism, generative AI, large language models, posthumanism.

1. Introduction

> Any technology that fundamentally reorients the ways humans exchange and interact with information must be taken seriously as a potential threat, or at least a disruptive variable, in the context of democratic life. It is therefore no surprise that fears over the rise of generative artificial intelligence (AI) and its potential usurpation of human agency and democratic self-rule continue to draw attention from the public and policymakers around the world.
>
> Wihbey 2024, p. 3

The expansion of *generative artificial intelligence* (GenAI), particularly large language models (LLMs), has spawned a range of theories and approaches aimed at defining whether these deep learning technologies genuinely reflect human intelligence. While some neuroscientists and linguists have suggested that LLM architectures accurately capture the mechanisms of human language production in the brain (and that, as a consequence, LLMs actually understand language in all its purposeful, agentive dimension (e.g. Tuckute et al., 2024), posthumanist and transhumanist thinkers have seized upon the fuzziness of LLM functioning to proclaim the impending obsolescence of humans as the sole conceptualizers of nature and our relationship to it. In this paper, I set out to refute these views, arguing that the reasoning behind such interpretations of GenAI and the projected role of humans is fundamentally flawed, marred by folk theorizing that encompasses "an inarticulate yet influential *science-before-testimony* picture" (Gerken, 2022, p. 1; emphasis in original).

For our purposes, ushering definitions that depict humans as mere performance-driven semiotic systems devoid of agency qualifies as a *before-testimony* conclusion. Specifically, I argue that human intelligence and consciousness cannot be reduced to mere computational processing for information integration, as proposed by theories like the *Information Integration Theory* (*IIT*, Tononi 2004). Instead, I suggest that consciousness arises from the holistic integration of perception, conceptualization, intentional agency, and self-modeling within biological systems. The present proposal seeks to contribute to the growing body of work in practical philosophy of science, a form of meta-science that focuses on the potential for science-making and theorizing to cause harm in the broadest sense (see also Shrader-Frechette 2014, p. 6). This inevitably leads me to define humans as an endangered species, thereby avoiding the endorsement of changes that benefit only certain segments, such as the ideological distinction between vulnerable individuals or communities, and privileged individuals, since "the survival of sentient biological life" (Hedlund & Persson 2024, p. 454) does not concern some individuals but all individuals.

Drawing upon an *Extended Humanist* perspective, I assert that the domain of human consciousness and agency is intrinsically tied to biological embodiment and the experiential interactions of conscious agents with their environment. The emphasis on mere information processing or disembodied existence, as envisioned by transhumanist visions of mind and digital consciousness endorsed by many scientists and philosophers, is viewed in this paper as a potential source of harm that fails to capture the role of human consciousness and agency in the construction of more inclusive and humane societies.

Similarly, posthumanist attributions of agency, consciousness, and meaning-making capabilities to machines and AI language generators neglect the fundamental role of biological embodiment and the holistic integration of perception, conceptualization, intentional agency, and self-modeling.

1.1. The brain-machine myth as an algorithmic ontology of the mind

1.1.1. The ghost's fingerprint

This paper endorses a *weak AI* position by which consciousness in machines can only be possible at the level of simulation of thought and sentience (e.g., Searle, 1980). However, the question as to whether machines have reached consciousness or will ever develop a sense of being in the world is useless (Levy, 2009), since what we can know about the conscious states of machines is defined by what we know about our own consciousness as a first-person phenomenal experience. In this sense, our comprehension of what a robot could do with thought and reason is mediated by our grasp of reality, our definition of objects and entities and what affordances thereof (in the Gibsonian sense) are salient enough and useful to us as part of the biological world.

To take a specific example, AI-powered robots need to be cognizant in the ways we define the term, that is, in terms of wakefulness and self-awareness, object identification and conceptualization, or how we deal with our surroundings in an agentive manner, through intention and purpose, through mind reading and making decisions on the basis of a survival imperative (see also Kinouchi & Mackin, 2018, p. 1, Torres-Martínez, 2024c). A diluted form of consciousness based on access to inner data and monitoring of processes (Block 1995) as a response to outward stimuli is, at present, the most feasible form of computational consciousness available to machines (Dahene et al., 2017). However, the problem in defining consciousness pertains inevitably to the possibility of having autonomous machinic agents capable of interacting with humans socially. These agents need to be capable of representing the world to themselves:

> So we see that representation means *to bring to oneself a picture of the world through bodily affordances, biological memory and incoming experience with that world*. In this sense, a picture of the world is both the result of perception and its further coupling with a chain of actions upon the world that is defined by the application of patterns of action-behavior in order to attain optimization (uncertainty-reduction) (Torres-Martínez 2024c).

More concretely, the prospect of machinic agency and the development of artificial consciousness depends critically on whether representations are connected merely with adaptive capabilities or constitute a fundamentally different type of unconscious agency. This alternative agency would deviate radically from biological systems by focusing on action execution within a performance range, rather than energy minimization for survival.

One of the primary limitations of the AI enterprise in developing conscious artificial agents lies in the constraints of hardware and available resources. These systems are inherently bound to exhaust their bodily and environmental resources while attempting to solve increasingly complex tasks through unsustainable processing schemas. Consequently, the pursuit of artificial consciousness reveals more about the developers' ambitions and misplaced expectations than about potential emergent machine intelligences, potentially leading to catastrophic consequences.

This perspective does not negate the fundamental human motivation behind artificial consciousness research: our intrinsic need to comprehend consciousness and intelligence. The creation of artificial intelligent agents essentially establishes a dialogue between the Human and the Other—the non-human conceptualizer. This dialogue has increasingly motivated investments in embodied forms of AI. For example, Reinforcement Learning (RL) exemplifies a nuanced approach to computational agency by which systems learn by pursuing goals through extended environmental interactions. As Butlin (2022, 2023) argues, these systems are fundamentally sensitive to their outputs' consequences, developing strategies that maximize cumulative rewards across multiple time steps.

The sophistication of RL becomes particularly

evident in complex decision-making scenarios. Researchers have developed computational models capable of simultaneously managing multiple reward functions, mimicking biological systems' ability to balance competing homeostatic drives. Studies by Keramati and Gutkin (2014), Juechems and Summerfield (2019), and Andersson et al. (2019) have explored various computational approaches to this challenge. Notably, Dulberg et al. (2023) proposed a model using multiple independent modules that learn to maximize distinct reward functions, with actions selected based on the highest total score.

The concept of embodiment, explored through output-input models or forward models, represents another critical dimension of artificial intelligence. These models enable systems to distinguish between environmental changes and self-generated sensory inputs—a hallmark of embodied intelligence.

In perceptual contexts, forward models allow systems to predict and differentiate the sensory effects of their own actions from external environmental changes. For motor control, they facilitate rapid and precise adjustments to complex effectors. However, current research suggests that while many AI systems employ forward models, few fully satisfy the embodiment criteria outlined in theoretical frameworks. A notable example is the system described by Friedrich et al. (2021), which utilized Kalman filtering (Todorov and Jordan 2002) to estimate a system's current state by integrating forward models with delayed sensory inputs. This approach demonstrates a sophisticated method of combining predictive models with real-time environmental interactions.

Despite these promising developments, the critical perspective previously articulated exposes fundamental limitations in the embodied artificial intelligence enterprise, particularly regarding resource constraints and representational challenges. As noted, artificial agents are constrained by materiality and potentially unsustainable processing schemas. Unlike biological systems that have evolved energy-efficient survival mechanisms, computational approaches to agency fail to adequately address the enormous energy requirements for creating adaptive systems capable of complex, open-ended problem-solving. This reveals that the computer metaphor has evolved within a conceptual vacuum that privileges specific pragmatist definitions of technology and *information-as-right-and-asset* at the expense of environmental constraints and broader systemic rights. This represents what I have termed *disembodied pragmatism*—a utilitarian mindset that relocates responsibility, accountability, and limitation into the realm of unbounded development of marketable information and informational goods and services, systematically undermining individual rights and the public good. The use of the American *First Amendment* for the protection of corporate speech serves as a quintessential example of this conceptual displacement.

> The justification for covering commercial speech under the First Amendment has shifted from protecting the interests of listeners, to protecting the interests of speakers, to valuing information for its own sake regardless of authorship or source. (Cortez and Sage 2023, 746)

It is no insight that over the last 60 years, the computer metaphor has nurtured the machinic-brain myth by which specific research designs have provided us with baroquely complex models of the mind based on the positing of arbitrary architectures with a weak *ontological basis*, but possessing *enhanced informational potential*. A paradigmatic set of beliefs backing up this mindset is the AI *transcendence hype* nurtured by Hollywoodian tropes by which "AI [is viewed] as the ultimate technology (and therefore the ultimate solution to all problems, i.e. pollution, the energy crisis, disease), yet at the same time the ultimate threat to humanity; the reduction of the individual to data and computation, and therefore the possibility of digital immortality" (Cave & Dihal, 2023, p. 4).

I will put this analysis on hold for the moment to concentrate on the impact the machinic myth has had on the current view of *language-as-code*. On the one hand, nativists like Chomsky (Chomsky et al., 2019) see the brain as a computer naturally wired to symbolic processes within a black box, which results in the generation of language assembled on the fly by an elusive linguistic competence encoded in our *linguistic genes*. On the other, usage-based cognitive linguists and constructionists (empiricists, e.g. Goldberg, 2019; Jackendoff, 2012) see the brain as a statistically-driven network of learnable symbolic representations fine-tuned to map and produce linguistic content.

Neuroscientists have taken heed of this theoretical claim in order to provide their research with definitions of language in line with some experimental paradigms (e.g., Kean et al., 2024). However, one problematic assumption of the computer metaphor in neuroscience is that language is conceptualized as consisting exclusively of symbols whose combinatorics can be localized in language-specific brain areas and identified during ad hoc tasks. This has resulted in the creation of a model of the linguistic mind that is insulated from other cognitive domains.

Moreover, neuroimaging technologies provide machinic representations of processes in the brain that cannot be interpreted as *reliable* renderings of the formation and distribution of concepts in the mind. This, of course, touches upon epistemological considerations that have consequences for the development of both neighboring and not so close fields of scientific endeavor. This also compromises the status of *fact* of the results provided by many studies considered as true and reliable within a single epistemic community: "[S]ome scientific ideas should be called 'facts', and they should be called 'facts' because they are true ideas (…)" (Vickers 2023, p. 2).

At this juncture, it becomes apparent that aligning with either a *scientific perspective*, as advocated by Massimi (2022), or a *future-proof science* stance, as proposed by Vickers (2023), offers limited assistance in defining true and reliable science. The reason lies in the inherent situatedness of scientific practice, which cannot independently justify the reliability of results accepted by a scientific community. This community operates within a framework of rule-following dynamics, wherein members endorse community-specific knowledge claims (Massimi 2022, p. 7).

Therefore, the reliability of scientific results, deemed as facts within a community, cannot simply be equated with the means by which these results are integrated into existing knowledge. This process often involves consistent adherence to conventions or institutional *rule-following* (Wittgenstein 1984 [1953]). Henceforth, then, I will stick with the argument that Wittgenstein's conclusion that rule interpretation is akin to rule postulation is flawed, since the reasons to do something according to a rule are not dependent on goals existing *prior to* the accomplishment of a "given task". The idea I am advancing here, then, is that an expected consequence of endorsing the rule-following paradox is that any institutional use of language, in the context of a form of life or language game, is always open to a hypothesis-driven reconfiguration of the content of the rule due to the intentional character of actions.

More concretely put, the content of rules is defined, not by *epistemic blindness,* and the systematic adherence to convention (as Wittgenstein suggested), but the predictive character of human cognition. It is worth pausing to see how this comes about. John McDowell (1992) suggests that, in order to prove a correct reading of convention, a rule has to be postulated that warrants meaning and understanding. This interpretation has a lot to be said for it, for example that, after having reached 1000 in a series I need not proceed to add 1002, since I may not want to proceed to add anything at all. In the same vein, the content of the underlying rule should not compel me to repeat the series *that way* in order to show that what I mean is in accord with a previous training. So we may take as proven that the crux here is that both the content of the rule and my grasp thereof are elements of the broader category of reasons which are not reducible to a "correct" or "incorrect" grasp of a rule within an epistemic community.

Seen in this light, both the Wittgensteinian and the McDowellian type of rule are isomorphic in that they establish a normative relation with a context created for it to be meaningful. Another way of putting it is to say that the context provides the norms that make my use of a rule meaningful and acceptable for a community. However, rules cannot be made to contain *behavior*, as reasons do not contain a norm to act in ways that privilege correctness over goals. Perhaps better put, agents engaging in a social use of language will always have the possibility to opt out conventional language usage in order to accrue epistemic capital.

Another consequence of this formulation is that the internal structure of rules cannot be framed in terms of a series of steps associated with family resemblance, that is, those stages entertaining an isomorphic relationship with the parts of an event during rule application.

The prevailing ontologies regarding language formation and processing in the brain cannot be simply derived from accepted methodological and epistemic principles, which suggest that highlighted brain regions serve as reliable evidence of ontological reality (e.g., Fedorenko, Ivanova & Regev, 2024), a mathematical "sketch of a model" (Mougenot & Matheson, 2024) that follows an epistemology that can be summarized as: 1) taking a concept from folk psychology (memory, attention, etc.), 2) identifying a suitable information processing theory, 3) correlating performance on a task, and 4) mapping this onto a brain region to explain the concept (Mougenot & Matheson, 2024). This approach can be considered folk theorizing, that is, the positing of constructs based on an implicit "science-before-testimony picture" (Gerken, 2022, p. 1) that wields significant yet inarticulate influence. A more bizarre assumption consists in applying folk theorizing to experimental outcome prediction, in the belief that neuroscience experiment prediction can be improved by LLMs. According to this assumption, "[i]f LLMs' predictions surpassed human experts, the practice of science and the pace of discovery would radically change." (Luo et al., 2024, p. 1). Now, is advancing the pace of discovery a realistic goal for a field defined solely by the application of technological metaphors to the study of a single organ through computer scans? Moreover, can the brain become a symbol definable by LLM data averaging? (See subsection 2.1)

In a similar vein, it is untenable to assert that a linguistic natural kind within the brain, poorly defined as "an ontologically meaningful grouping of brain areas" (Tuckute et al. 2024, 282), can be grounded solely on the traditional encoding-decoding process of generating meaningful symbolic strings. Crucially, the deficiency-fallibility bias backing up symbolic output-oriented task design characterizes human "performers" as "*fallible* but boundedly *rational* agents who can in principle *eliminate their ignorance* as long as the *task* lies within *cognitively allowed* applications of *inference rules*" (Solaki, 2022, p. 543; my emphasis). As the wording of the previous passage shows, there is an increasingly ubiquitous irony in current algorithmic-driven epistemologies: whereas the Turing test aimed at finding out whether "someone [could] tell accurately whether they [were] communicating with an AI or a human in a decontextualized interaction" (Youssef et al., 2024, p. 6, citing Nov, 2023), the effort is now on determining the extent to which humans can form representations allowing them to communicate efficiently with machines through a reduced set of signs defined a priori by the training of a language model.

As previously indicated, this approach often incorporates unfounded assumptions about the psychological reality of signs-as-action. For instance, defining language as consisting of form-function pairings ("constructions", as proposed by Goldberg, 1995, 2019) imposes a particular framework for analyzing linguistic data during the online monitoring of brain activity.

Moreover, the use of non-invasive neuroimaging techniques such as electroencephalography (EEG) and magneto encephalography (MEG), used to measure the electrical or magnetic activity of the brain, respectively, has provided tools to reinforce a narrative whereby whatever displays either neurochemical electrical or magnetic activity within a response-to-stimuli timeframe during specific tasks attests unequivocally the existence of a physical anchoring for language and the concepts encoded by it. Therefore, the idea of *functional localizers* (Tuckute et al., 2024), that is, a set of stimuli used to detect where a specific linguistic behavior "resides", suffers from a lack of conceptual support other than the formulation of stimulus-matching conditions that make no distinction between signals and signs during language representation (that is, $Language = \Sigma(Symbols, Rules)$). The ensuing dehumanization of the brain seeks to reduce the complexity of the embodied components of language production and its connection with the identification of conscious states, event recognition-partition, and the agentive character of linguistic actions. According to this reading, the linguistic brain possesses a type of predictable internal variation definable by the amount of observable activity in specific brain areas displaying a distinct pattern of interconnection, which results in the positing of a *ghost fingerprint* (an index) by which a non-reducible natural kind can be identified as an ontological object in its own right. This characterization brings to the fore both an essentialist (the *how*) and an intrinsic (the *what*) definition of the brain as a construct existing on a single layer of reality expressed by one-dimensional propositions, which can be summarized as follows:

1)

$$\text{Brain } 0_{\text{Language State}} = \text{stimulus} \rightarrow \text{Language} \neq \Sigma(\text{Brain\_Areas})$$

This, I oppose using the formula:

2)

$$\text{Brain} \neq f(\text{Language\_State} \mid \text{Stimulus} \rightarrow \text{Language-Specific\_Network})$$

This formula captures the following points:

1. The brain is not a function of the language state, where the language state is determined by a stimulus leading to the activation of a language-specific network in the brain.

2. The Language ≠ Σ(Brain_Areas) formula reflects the idea that language cannot be reduced to a sum of specific brain areas.

3. The use of the conditional probability symbol "|" emphasizes that the language state is conditionally dependent on the stimulus and the assumed language-specific network, which is rejected in this paper.

4. The negation symbol "≠" indicates that the brain is not adequately represented by this indexical model of language processing, where a stimulus activates a language-specific network to determine the language state.

Since one of the attributes of natural kinds is *iconicity*, that is, they are susceptible to being perceived as distinct signs resembling what they stand for, it is not possible to announce the emergence of a brain-based natural kind based on indexical information only. Hence, what is intrinsic and essential in a natural kind cannot be defined through reaction or contiguity, as the reasons why something happens or is in *that* particular way and not another, remain underrepresented in the mind of the interpreter-conceptualizer. This can be summarized by the following sets and functions:

U = Universal set of all possible signs

I = Set of iconic signs (resembling the object they represent)

X = Set of indexical signs (indicating the object they represent through a causal or physical connection)

N = Set of natural kinds

B = Set of brain-based natural kinds

S = Set of signs used to announce the emergence of a brain-based natural kind

We can represent the disjoint nature of iconic signs and indexical signs using set operations:

$$I \cap X = \emptyset$$

Additionally, we can define a function $f: U \rightarrow [0, 1]$ that represents the degree of iconicity of a sign, where 0 indicates no iconicity, and 1 indicates a perfect iconic representation. We can define this function as:

$$f(s) = 1, \text{ if } s \in I$$

$$f(s) = 0, \text{ if } s \in X$$

To represent the essential nature of iconic signs in representing natural kinds, we can define a function g: N ⊠ [0, 1] that measures the degree of iconicity required for a sign to represent a natural kind:

$$g(n) = \alpha, \text{ where } \alpha \in (0, 1]$$

The value of α represents the minimum degree of iconicity required for a sign to adequately represent a natural kind. We can then establish the following condition:

$$\forall n \in N, \exists s \in U: f(s) \geq g(n)$$

This condition states that for every natural kind *n*, there exists at least one sign s in the universal set U whose degree of iconicity *f(s)* is greater than or equal to the required degree of iconicity *g(n)* for representing that natural kind. Furthermore, we can represent the limitation of using only indexical signs to announce the emergence of a brain-based natural kind by defining a function $h: S \rightarrow [0, 1]$ that measures the degree of indexicality of the set of signs *S* used for this purpose:

$$h(S) = (1 / |S|)\ \Sigma\ (1 - f(s)), \text{ for all } s \in S$$

Here, $|S|$ represents the cardinality (number of elements) of the set *S*, and the function *h(S)* calculates the average degree of indexicality of the signs in S by summing up the complement of the iconicity values $(1 - f(s))$ for each sign s in S and dividing by the total number of signs.

We can then establish the following condition:

$$\forall b \in B, h(S) < \beta$$

This condition states that for every brain-based natural kind *b*, the degree of indexicality *h(S)* of the set of signs *S* used to announce its emergence must be less than a threshold value β, where $\beta \in [0, 1)$.

In conclusion, we should resist the temptation to conjure up a Laplacean demon possessing a set of "compact truths" from which all other truths can be inferred (contra Chalmers 2014, xiv) and then impose this framework onto a theory of a linguistic natural kind in the brain.

1.1.2. The "predictive" black box

Embodiment theory is gaining traction in contemporary research (e.g., Lux et al., 2021). The discussion is often framed around whether concepts are represented in modality-specific formats, as demonstrated by effects like the *action-sentence congruency effect* (ACE; Glenberg & Kaschak 2002), or whether perceptual reenactment is unnecessary for conceptualization (e.g., Mahon, 2015a; Vannuscorps et al., 2021). A working definition of embodiment for a theory of mind can be formulated as the process by which cognizers integrate external object-entity affordance configurations (external cognition) with the internal processing of somatosensory signals (internal cognition) to form approximate representations of the world. This perspective challenges the notion that perception generates its own images, emphasizing instead that "it is the relevant tract of the environment that is present to consciousness, not an image of it" (McDowell 2002, p. 455). For example, consider frogs. Neuroscientists might claim that the frog's eyes convey specific information to its brain about an object in motion, but, as McDowell (2002) argues, this "telling" does not originate internally. Instead, the environment itself, through its affordances, provides the frog with actionable information, facilitated by the frog's perceptual apparatus (McDowell, 2002, p. 450). Here, I posit that our understanding of the world is mediated by the body via *multiple modalities*—such as proprioception, interoception, and exteroception—and further structured by somatosensory maps, which deliver sensory feedback to the brain (a process described here as *extended cognition*). This argument counters the view that perceptual information is merely re-accessed in modality-specific systems or brain regions (e.g., Barsalou, 1982, cited in Machery, 2010). Instead, cognition is framed as a distributed process involving the body and the environment.

On the other hand, the independence of thought from perception, as suggested by Mahon (2015a, 2015b, 2015c), assumes that thought resides solely in the brain. However, this paper supports the thesis that online reenactment of perceptually driven representations arises from action-perception loops. These loops, however, should not be simplistically reduced to limb-action matching (e.g., Mahon 2008). Instead, cognition occurs beyond the brain, extending to the body and the environment. From this perspective, the distinction between concrete and abstract concepts dissolves, as the form and content of representations are accessed as they are. The framework proposed here prioritizes biological processes over psychological ones, viewing cognition as involving entities external to the agent's body within its local environment (Sims and Kiverstein 2021). Consequently, purely symbolic concepts are untenable, as they possess semiotic properties—iconic and indexical (see Torres-Martinez, 2022b)—that do not necessitate specific action-related mechanisms to achieve representation or systemic activity (contra Mahon and Hickok, 2016).

An even more radical conceptualization of the disembodied brain has been proposed by theoretical neuroscientists, in the context of the *free-energy principle*

(Friston 2010), who project mathematical models onto the biological function of the brain in order to create a functional abstraction that, nonetheless, fails to create an accurate representation of the role of language in the construction of reality. In this context, the notion of linguistic competence has been extended to computational models of the brain consisting of hierarchical, bottom-up processes whereby the "wetware" is deemed to work as an energy minimization machine (Murphy, Holmes and Friston 2024), with language being conceptualized as a code separated from its agentive dimension: "[E]mbracing the traditional distinction between competence and performance, our focus will be on the former (the mental formatting and generation of linguistic structure), and not on the range of complex cognitive processes that enter into the use of language in a specific context" (Murphy et al., 2024, p. 8).

Clearly, the authors introduce a model of language generation that reduces the cognitive processes involved in producing language to efficiency-led operations mapped onto an idealized brain-kind, which can hardly be made to correspond to actual brain functioning, as defined by neuroscience. Indeed, computational models of language processing do not necessarily overlap with the brain's chemical and electrical reactions associated with identified language regions (Tuckute et al., 2024).

While I would be the last to negate the benefits of proposing a predictive, uncertainty-minimization model of the brain (see Torres-Martínez, 2023a; 2024a,b,c), it is also true that human language cannot be defined by merging symbolic syntax with Bayesian models of the brain.

What we get is a self-referential mathematical model more concerned with maintaining internal theoretical coherence than providing evidence of its utility as a basis for the identification of ontological objects and natural kinds in the brain-body interface.

As I have stated elsewhere (Torres-Martínez, 2024a,c), a more refined reading, informed by active inference (*AIF*, Friston 2009, 2010, see Section 3), posits that language, as part of a biological system, serves to link an organism's need to negotiate uncertainty with its goal-directed actions impacting the world. In this sense, a variation of Murphy et al's. formalism (2024, 5) states that the expected free energy $G(\pi)$ for a sequence of actions (policy) π over time t is given by:

$$G(\pi) = \int_t G(\pi, \tau)$$

Where $G(\pi, \tau)$ at the current time step τ is:

$$G(\pi, \tau) = E_Q[\ln Q(\tilde{s}|\pi) - \ln Q(\tilde{s}_\tau|o_\tau, \pi)] - E_Q[\ln P(\tilde{o}_\tau)]$$

This can be rearranged as:

$$G(\pi, \tau) = E_Q[\ln Q(o_\tau|\pi) - \ln Q(o_\tau|\tilde{s}_\tau, \pi)] - E_Q[\ln P(\tilde{o}_\tau)]$$

The components are:

1. Negative Mutual Information: $E_Q[\ln Q(\tilde{s}|\pi) - \ln Q(\tilde{s}_\tau|o_\tau, \pi)]$ represents the reduction in uncertainty about the state $\tilde{s}$ by taking actions π compared to the expected uncertainty given observations o_τ and actions π.

2. Expected Log Evidence: $E_Q[\ln P(\tilde{o}_\tau)]$ is the expectation of new evidence or observations $\tilde{o}_\tau$ based on the current actions π and observations o_τ.

3. Negative Epistemic Value: $E_Q[\ln Q(o_\tau|\pi) - \ln Q(o_\tau|\tilde{s}_\tau, \pi)]$ quantifies the reduction in uncertainty about outcomes o_τ through actions π versus the expected uncertainty given the state $\tilde{s}_\tau$ and actions π.

4. Extrinsic Value: $E_Q[\ln P(\tilde{o}_\tau)]$ represents the expected benefit or utility of outcomes $\tilde{o}_\tau$, regardless of the current state and actions.

We can thus assert that the expected free energy balances the extrinsic value (desirability of outcomes) and epistemic value (information gain) when selecting actions. Actions that minimize expected free energy are favored, as they lead to desired outcomes while reducing *uncertainty*.

Seen through the lens of a non-artificial modeling of language construction with AIF, an organism can only maintain its internal homeostatic integrity when capable of detecting and reducing the effects of environmental disturbances. In other words, when the organism's internal states are attuned to the external conditions shaping its interactions with the world. A statistical boundary (the Markov blanket) interfaces incoming sensory signals, providing the system with information about external conditions. Perceptual surprise is thus a crucial evolutionary motivator, as the degree of acquaintance with the unknown is what ensures survival.

From this perspective, emergent human linguistic behavior is structured around recurrent hypothesis-driven semiotic models (termed herein as *Constructional Attachment Patterns* or CAPs, Torres-Martínez, 2018a,b, 2019, 2020, 2021a,b,c, 2022a,b) , that is, hypothesis-driven semiotic relations among constructions that facilitate an embodied, perceptually-driven type of reality reconstruction (see Torres-Martínez, 2022a). For example, CAPs are integral to broader cognitive processes that enable individuals to evaluate, predict, or hypothesize about events involving others' decisions, attitudes, beliefs, and behaviors. These include mechanisms like *action recognition* (e.g., Ferstl et al., 2017), which involves identifying and interpreting others' actions, and the reflection of specific mental simulations in language. For instance, simulations of speed are mirrored in linguistic structures (see Speed & Vigliocco, 2014; Pan, Liang & Li, 2024), as is the processing of abstract action sentences (e.g. Schaller et al., 2016; Balduin-Philips et al., 2021). Additionally, CAPs support cognitive tasks like *spatial perspective taking*, which involves adopting someone else's viewpoint to understand spatial relationships (e.g. Kessler & Thompson, 2014; He, Chrastil & Hegarty, 2022). These examples illustrate how CAPs function within the complex interplay of cognitive, perceptual, and linguistic processes, emphasizing their role in bridging mental simulations and linguistic expression.

CAP selection is governed by "agentive imperatives" rather than purely statistical mappings, thereby creating meaningful categories. This process is summarized by the formula below (see also Torres-Martínez, 2021a,b): Ultimately, specific CAP selections are employed to define, shape, and guide an agent's actions in the world. In simpler terms, the association of events—bridging two experiential domains through event segmentation and sequencing—must meet two criteria: 1) it should be retrievable through a coherent and non-conflicting combination of CAPs, and 2) it must be experientially accessible to both the conceptualizing agent and the intended addressee. Relying on CAPs (Constructions as Attachment Patterns) rather than static signs provides significant advantages for understanding language. CAPs align with a dynamic view of the mind, where reality is constructed through the interplay of prior knowledge and new sensory input (see Torres-Martínez, 2020, 2021b, c, 2022a). This perspective supports the idea that language is not a fixed code but a flexible, adaptive tool shaped by users to navigate and interpret their experiences. CAPs seamlessly integrate embodied cognition—how physical and perceptual experiences shape thought—with intentionality, offering a multidimensional model of language that emerges from the interaction of speaker goals, contextual factors, and linguistic materiality. Thus, language is treated as a semiotic system actively molded by speakers, rather than as a static repository of symbols. CAPs show that language is a user-driven system, enabling speakers to creatively combine diverse signals—ranging from phonetic and tonal features to syntactic structures—into coherent, context-sensitive utterances. This adaptability reflects the speaker's ability to resolve uncertainty by tailoring language to specific communicative needs.

In this sense, the idea that language emerges from statistically relevant associations between objects/ entities (natural kinds) being noticed by infants, during their interaction with caregivers, and its further coupling with specific argument structure constructions (syntactic templates with slots for lexical content), provides a very poor reading of language as a cognitive tool that is not governed solely by symbolic permutations moving on a continua of entrenchment and weakening associations. This computational reductionism of language has given a weak edge to usage-based linguists to postulate a possible link between human language construction and AI-generated language. (e.g., Beuls & Van Eecke, 2024)

On the contrary, human agents are defined by a

type of *bodily awareness* – an umbrella term to refer to the representation of Self in the world through the conjunction of perceptual states, their conceptualization, and the projection of beliefs onto specific events in a context of use. In essence, this viewpoint positions language as arising from an organism's need to reduce uncertainty and sustain itself by accurately modeling its environment, with linguistic constructions emerging from embodied, agent-driven processes rather than purely statistical learning over data. In this sense, it is not possible for current artificial models of language generation to extract meaning from situated interaction with humans simply because human experience is not limited to the confines of an incremental acquisition-verification loop of linguistic data in order to attain a communicative goal. This of course, contradicts the argument introduced by Turing and others, and reinforced constantly by computationalists (e.g., Beuls & Van Eecke, 2024, 2025; Weissweiler et al., 2023), by which simulation of sentience and consciousness amounts to actual conscious states in a language game:

> We hope, however, to have convinced the reader that a better integration of the situated, communicative, and interactional aspects of human linguistic communication constitutes the key to overcoming the limitations of current LLMs, and that more faithfully modeling the situated communicative interactions through which humans acquire their native languages provides a promising path towards more human-like language processing in machines. (Beuls & Van Eecke 2024, p. 28).

2. The risks of folk theorizing: Extended humanism or posthumanism?

In this section, I lay out the foundations for the conceptualization of consciousness and first-person experience. No claim is made to completeness or definiteness in this theoretical quest. I am content simply with delimiting the contours of an emerging field of endeavor that, on the one hand, shows how far we have gone in our quest for the destruction of the human and what being human reveals about the substance of the world, and, on the other hand, unmasks the lack of consistency in our current knowledge of our very nature and purpose in the universe.In this section, I lay out the foundations for the conceptualization of consciousness and first-person experience. No claim is made to completeness or definiteness in this theoretical quest. I am content simply with delimiting the contours of an emerging field of endeavor that, on the one hand, shows how far we have gone in our quest for the destruction of the human and what being human reveals about the substance of the world, and, on the other hand, unmasks the lack of consistency in our current knowledge of our very nature and purpose in the universe.

These reflections will revolve around the need to think of the human as a manifestation of a distinct set of properties that separates us from the non-human, especially emerging artificial agents. In order to dispel possible misunderstandings at the outset, I am compelled to say that my views of humanness and identity are not rooted in a Jungian "development of the I" (Harding 2020[1965]), but in the concept of the *extended human*, that is, an open biological organism bestowed with consciousness and aware of their deep connections with the natural world and other sentient and non-sentient beings. Extended humanism seeks to downplay the underlying tension plaguing discussions of AI/

deep learning advancement usually split into views defined by the overconfident tone of nurture-favoring machinic neo-empiricism (privileging an unbound algorithmic narrative of machines "on the verge of achieving escape velocity into world-spanning superintelligence (…) (Buckner, 2023, p. 4), or the *nativist* (also referred as *innatist*) position by which rational thinking can only be achieved by humans thanks to "the human mind's innate startup software" (Buckner, 2023, p. 4).

My project also seeks to dismantle the dehumanizing rhetoric of (empiricist-driven) posthumanism, a program that, as suggested in section 1, extends over several fields through a parsimonious deconstruction of the Self as a means to avoid (nativist) anthropocentrism and, more particularly, monolithic forms of Western-centric "we" narratives (nos majestatis, as the Romans called it) (Massimi, 2022, p. 366). As we will see, however, the result has been the emergence of an antihumanist discourse, fueled by beliefs about 'virtuous' and 'vicious' theory building (West, 2024) that relativizes the human body and mind (for instance, scientists are often pigeonholed on the basis of their adherence to politically correct discourses, rather than being recognized for the soundness of their contributions), which negates the complex role of humans as a species in need of broader strategies to defeating the "darker side of [their] success" (Desmond & Ramsey 2023, p. 3). These include, as noted previously, broadening the scope of theoretical philosophy of science so as to include more practical questions such as why humans uncritical overreliance on technological advancement may cause harm to all of us, not only particular vulnerable groups (a definition that is in itself biased towards the "sustainable-development" of politically correct narratives through AI ethics, see Curto et al., 2024).

Therefore, in the context of AI-generated language and human-machine interaction, many philosophers and researchers have positioned pretrained transformers and chatbots like ChatGPT as distinct agents capable of creating culture and meaning.[1] This extended *engineering empiricism* "draws oxygen from a simplistic understanding of the relationship between the successes of deep learning systems and the way that humans and animals actually solve problems." (Buckner, 2023, pp. 3-4). As a result, meaning is treated in very a loose manner. For example, it has been argued that

1 The language generation process of ChatGPT begins with tokenizing input text XX into a sequence of tokens X = {x1, x2, ..., xn}. Each token xi is transformed into a high-dimensional vector representation ei through an embedding function E(xi) = ei
The model employs a Transformer-based architecture with layers defined by a generalized transformation:

$$\text{Layeri(input)} = \text{LayerNorm(MultiHeadAttention(input) + input)}$$

The attention mechanism, a key component, is calculated using the equation:

$$\text{Attention}(q,k,v) = \text{softmax}(q \cdot kT / \sqrt{dk}) \cdot v$$

Where q, k, and v represent query, key, and value matrices, respectively. The softmax function with scaling by √dk helps manage the dot product's magnitude.
During training, the model minimizes the cross-entropy loss:

$$L(Y,\hat{Y}) = -\Sigma i \ yi \log(\hat{y}i)$$

This loss function measures the divergence between predicted and actual token probabilities. The model can be fine-tuned on domain-specific data by updating parameters using gradient descent:

$$\theta new = \theta old - \alpha \cdot \nabla \theta L$$

Where α represents the learning rate and ∇θL is the gradient of the loss with respect to the model parameters. In the decoding phase, the next token x̂t+1 is predicted by maximizing the conditional probability:

$$\hat{x}t+1 = \text{argmax\_x } P(x \mid x1, x2, ..., xt)$$

This means selecting the token that maximizes the probability given the previous tokens.
Finally, the generated text undergoes post-processing steps like detokenization, filtering, and formatting to produce a coherent and polished output.

machines that produce some kind of autonomous meaning are *agents* and possess consciousness. The claim goes so far as to assign meaning-making properties to processes that, by virtue of an empiricist heuristic, emerge as embodied cognitive entities with broadened forms of thought, "depending for its specificities on the embodied form enacting it" (Hayles, 1999, p. xiv). This conflation of sign-making processes and output, which does not necessarily contain meaning, has served as a conceptual tool for elevating all sorts of sign users to the status of conscious beings, under the belief that all it takes to be an agent is to process information to complete a task.

The implication of this is that a bedrock condition for consciousness to exist and provide context for meaning is the possibility of an entity to expand its cognitive grasp, offload intentional content, and predict what-will-be-the-case within an event that needs to be represented as an actual happening having bearings on a cognizer's grasp of existence. Sign production is thus the result of what is known by the cognizer in terms of purposeful meaning-making and not what is generated in the form of signs through natural or engineered computations. Inevitably, this opposes unconstrained definitions of cognition as "a process of interpreting information in contexts that connect it with meaning" (Hayles, 2017, p. 22; cited in Hayles, 2024, p. 31; Hayles, 2021, p. 6). The reason is that representation is not a homogeneous process geared by "pure" signs directed to and emerging from a species-specific Umwelt (see Torres-Martínez 2023a, b, c), but a nuanced blending of property identification, category assignation, and intentional contextualization for the emergence of a distinct sign system providing an organism with identity.

The semiotic juggling that conflates information processing with consciousness is also predicated upon the notion of *language-as-performance,* by which LLMs are elevated to the status of "the best predictive models of human language representations at the resolution of data [neuroscientists] have access to" (Tuckute et al., 2024, 285), an idea that has been integrated into ethnographic, antihumanist (posthumanist) research agendas seeking to ascribe an ontological status to pretrained transformers in the belief that these language generators are *authentic* language creators (e.g. Demuro & Gurney, p. 2024).

According to this reading, such models actively contribute to the cultural construction of knowledge through unique "reasoning processes" that, nonetheless, are distinct from those of humans. This assertion stems from a general rejection of any "universalist" definition of language which is associated with an anthropocentric humanist tradition, and instead views language as a manifestation determined by the *performance* of all sorts of agents through *communication-as-action*. Consequently, pretrained language models are seen not just as performers through language, but as active participants in its creation, thereby shaping events where language unfolds through both performance and competence, which "relativize[s] the human by coupling it to some other order of being" (Clarke 2008, 3).

Fittingly, human language is depicted as a process of assembling symbols within a grid of semiotic relations determined by the shifting shapes and affordances of specific cognitive technologies and objects (e.g., Hayles, 2021), with speakers functioning as participants bound by the rules of extant signhood preservation practices (language games) within a self-sealing *field event* (Hayles 2018). As a result, human language is portrayed as a residual aspect of interaction, lacking any inherent transcendental qualities (humanness), which redefines LLM language generation as an integral component of authentic "encounters" where participants, both human and non-human, share a sense of experiential adjustment (e.g., Demuro & Gurney, 2024; Dynel, 2023). This shift obliterates any claim to a construction of the Self-as-symbol through an emphasis on a redefinition of power relations that replace (anthropocentric) universalism with a focus on addressing gender inequality.

> My anti-humanism leads me to object to the unitary subject of Humanism, including its socialist variables, and to replace it with a more complex and relational subject framed by embodiment, sexuality, affectivity, empathy and desire as core qualities (Braidotti 2013: 26).

And yet, since human agency involves purposeful, goal-oriented action aimed at both reducing uncertain states and constructing preferred models of interaction with the environment, human beings, as distinct biological systems, participate in a continuum of relations with the world and other species that goes beyond denaturalized claims to power redistribution through the effacement of the matter-energy boundary. Under this view, it is not possible to posit an essentialist (though undefined) semiotic substance ("language") that can be reduced to underrepresented linguistic practices propelled by language use ("languaging"), definable a priori by a particular set of beliefs in the mind of an interpreter-anthropologist (e.g., Demuro & Gurney, p. 2023).

The primary contradiction in the posthumanist reading lies in the fact that even the most contingent set of relations is driven by biological imperatives that cannot be relativized through the postulation of indefinite intrinsic properties tied to a material network of relations:

For posthuman theory, the subject is a transversal entity, fully immersed in and immanent to a network of non-human (animal, vegetable, viral) relations (Braidotti, 2013, p. 193).

As we can see, the passage offers a chain of rhetorical devices whereby the unsupported negation of humanness, through the medium of "stylistic use of prose to evade clearly stating one's argument" (Rickabaugh & Moreland, 2023, p. 5, citing Searle 1992, p. 9), is dissolved in a definition that reifies an amorphous *substance of things* that denies the pivotal role of the human in the comprehension of the relations we are part of. Braidotti's wording also points to an underrepresentation of the intrinsicality of the human in the relations she attempts at describing, since, as I have shown elsewhere (Torres-Martínez, 2024d,e), the property of being intrinsic (the "what") cannot be separated from the deployment of the essential (the "how").

This new materialist reading of the Human (Coole & Frost, 2010), assumes that, since humans and machines are material renditions of broader processes of energy exchange and conservation, consciousness can reside in mathematically feasible recipients other than biological entities. Furthermore, agency is not a property of sentient beings but an affordance of phenomena defining cause-effect events (landslides, storms, hurricanes, and so on). On this reading, proponents of mathematical theories view LLMs as a means to accurately predict how meaning can be extracted from symbols and mapped onto the internal representations and concepts possessed by speakers immersed in an idealized state of *unity*, which is "often thought of as part of a descriptive theory of ideal rational agents, not of real agents" (Kindermann & Onofri, 2021, p. 2). A prime example of this mindset is the attribution of essential properties to chatbots, as Dynel (2023) states:

> Since ChatGPT can adapt its responses to match various discourse styles and registers, users can witness and understand how the model adjusts its language based on the input provided (p. 122).

In this passage, the author ascribes essential properties— qualities that allow something to behave or act towards a goal— to the chatbot. This suggests that its adjustments through training and enhanced processing capabilities create a form of *ontological dependence* between the user's needs and the prompt-driven event governing the generation of "polite" language. According to Dynel, this ontological dependence actually mirrors the inherent naturalistic linguistic interactions between human agents. The point to notice, however, is that the human perception of AI-generated language requires an omission of the layers of abstraction inherent in computational artifacts. These layers include, among others, the designer's intentions for the solution formulated to address a given problem. Moreover, it is crucial to recognize that displaying polite behavior (the manner in which someone behaves politely) does not inherently equate to possessing an *intrinsic quality* of being polite (the inherent quality that makes someone behave politely). Consequently, the suggestion that chatbot-human interaction entails a degree of naturalistic alignment of intentions, involving meaning negotiation and pragmatic adjustment, becomes implausible when only one participant, the human agent, possesses awareness of their conscious states enabling them to plan beyond the immediate chatting event (the AI's behavior is guided solely by the prompt imposed by the human). Clearly, [t]here is no *one* at the other end for the human to communicate with. Human–computer interaction is not inter*personal* communication" (Duncker, 2020,p. 97).

The type of conceptual relativism and attitude

revisionism proposed by posthumanism had been previously framed by Turkle (2005, p. 59) in terms of an "adulthood" bias by which grown-up humans attribute consciousness only to organic entities:

> The children take a different view. The idea of an artificial consciousness is not what impresses them. They may be the first generation to grow up with such a radical split between the concepts of consciousness and life, the first generation to believe that human beings are not alone as aware intelligences. The child's splitting of consciousness and life may be a case where instead of thinking in terms of adult ideas "filtering down" to children, it makes more sense to think of children's resolutions prefiguring new positions for the computer culture to come.

Indeed, the postulation of humans, "a strange product of evolution that can perceive itself and its fellow humans as paradoxical" (Rosendahl, 2013, p. 224), as self-symbols striving for biological priority and persistence, provides a solid basis for our definition of language as a cognitive tool that cannot be fully emulated by artificial systems of language generation.

Such radical embodied position goes against the grain of a growing realization among chatbot users that sentience, and even a soul-like otherness, is a by-product of our interaction and our conditioned reactions to chatbot's apparent empathy and well-calibrated responses. Sentience attribution is thus the result of a psychological projection onto an artificial entity that "enlivens the robots, even as the people in their presence are enlivened, sensing themselves in a relationship" (Turkle, 2010, p. 85).

2.1. Consciousness attribution leads to output averageness

The problem of sentience attribution in the study of human perception, regarding both AI-generated and human-constructed content, is that it is not evident that creativity attribution to LLMs as intelligent tools could eventually be used as a means to postulate a human flawed heuristic (e.g., Porter & Machery, 2024). In other words, it is not the case that human heuristic judgments could be compacted in a single-layered experiential proposition that excludes other experiences (e.g. *Information Integration Theory (ITT) 3.0*, Oizumi et al., 2024, 3, see section 3 below) and, hence, susceptible to being subsumed to the affordances of LLM performance, an idea that is central to ITT 4.0's formulation:

> By the intrinsicality postulate, the Φ-structure [the structure of the amount of consciousness] of a complex depends on the causal interactions between system subsets, not on the system's interaction with its environment (except for the role of the environment in triggering specific system states). In general, different physical systems with different internal causal structure may perform the same input–output functions (Albantakis et al. 2024, 36).

The reason is that, as a form of inference, heuristic (clue driven) decisions do not depend on a bias towards *mathematical simplicity*, that is, towards mechanized, meaningless habits within a closed physical system, since the nature of pattern finding for complex biological systems is reliant on a bottom-up identification of those elements (extracted from environmental signaling) that can provide a framework for the postulation of *emergent minimal working systems* (conceptual systems providing agents with a sense of consciousness-as-action). In terms of creativeness, the implication is that AI-generated output provides a minimal working system that accommodates clue-based inference without there being a need to *reduce world complexity*.

Thus, the mixing of *creativity perception, content generation, and machinic representation* as a graded dimension of attribution (the assignation of properties to a tool), has led some to conclude, for example, that "people prefer AI-generated poetry to human-authored poetry, consistently rating AI-generated poems more highly than the poems of well-known poets across a variety of qualitative factors" (Porter & Machery, 2024, p. 9). Notwithstanding, the preference for AI generated content among chatbot users does not imply poor heuristics, but a reduced investment in context analysis through enhanced attribution of precision and accuracy to AI technologies for precision leveraging and human outperformance (Jablonka et al., 2023). In other words, the use of empty signs results in blind sign deployment: "By using signs emptily, we do not even terminate in the consciousness of whatever it signifies anymore, because we use the habitual sign itself habitually, we do not invest mental activity into bringing the signified into focus, since we 'know' what it means." (Arnold, 2022, pp. 216-217)

Clearly, the conflation of human inference with AI content generation efficiency introduces a major risk, namely the elevation of *output averageness* through data -AI output tends to a mediocre flattening of reality, as is the case in medicine (Clusman et al., 2023; Schmidgall et al., 2024) - as the new direction in all conceivable fields of human action, science included. For example, enthusiasts of *enhanced* science-making through AI processing, assert, quite unashamedly, that the human mind should eventually become an outgrowth of AI technologies as a means to attain Noble-prize level influence and recognition through "top-level science" pattern definition, thereby positioning AI technologies as a solvent for human error and bias by means of a "Turing grand challenge" (Kitano, 2021, p. 1): "a grand challenge to develop AI systems that can make significant scientific discoveries that can outperform the best human scientist (…)".

However, the reduction of science to technology and "innovation" by way of perplexity values (that is, AI performance values leading to "efficient algorithmic discipline"), announces a world where bad science will become instrumental to a halt in human problem solving and insight. The stated goal of endorsing AI-powered automation as a standard for experiment design, data analysis, information collection, and results interpretation (e.g. Boiko et al., 2023), is thus to increase output optimization, rather than true scientific development. But such advancements come with a cost, namely that LLM deployment is operated on the premise that users "will do and fill the gaps", thereby resulting in unsupervised applications with weak LLM uncertainty quantification (see Catak & Kuzlu, p. 2024): "There remains a possibility for users to elicit fundamentally novel behaviors from LLMs, unforeseen by developers, through causal reasoning processes akin to those observed in GPT-3, defying current evaluation or analytical methodologies." (Kumar, 2024, p. 39).

As we will see, however, the emergence of "disembodied subjectivities", including not only the artificial, but the projection and ultimate extension of human cognitive states onto devices and non-human *others*, reinforces the idea that biological agency, driven by phylogenetic rules for survival and semiotic persistence, can guarantee a relevant place for our species in a world dominated by performative actants. To note, my use of "persistence" does not seek to position a code-driven idea of biological reality through heredity only.

As it will become evident in the next pages, my intention is to delineate the conditions for the emergence of an extended organic species evolving within a context of predictive semiosis not restricted to the mere passing of information through DNA-based heredity. That said, the idea of humans transcending their mortal organic bodies (transhumanism) as a means to outpace non-organic performers is replaced in this paper by a type of human that does not surrender their symbolic self to the promise of enhanced signhood through a machinic turn.

Nor is it suggested that the road ahead for "the Human" implies overcoming the old self to become more tethered to a shifting parcel of reality defined by an endless loop of efficacy and loose sense of progression through disruptive "encounters". Biological signhood is thus the result of pushing aside both the organic and the technological myth—cherished by some as a means "to compose enactment" (Hayles, 1999, 196; Hayles, 2012)—that separates us from the world of the living and non-living. In this sense, the idea of language as a semiotic tool for self-preservation and the construction of preferred futures entails an irreducible sense of being in the world that cannot be emulated by machines, since "life and sentience are coterminous. Life is based on cells, biomolecular entities of immense complexity. The *living* element, the biological component, is the critical factor." (Reber et al., 2023, p. xiv; emphasis in original).

The present conceptualization also highlights the idea that being embodied does not simply mean to act through and with a physical body but entails the existence of a sense of boundary that provides us with bodily identity.

This is precisely one of the problems of non-embodied approaches that attempt to fuse the human body with the objects profiled by our extended cognitive affordances (e.g. Hayles, 2012). Therefore, in postulating markers of cognitive utility (but, paradoxically, of semiotic dispersion and discontinuity) and usability as internalized "nonconscious" extensions of the cognizer, "the nonconscious cognitive assemblages through which these distributed cognitive systems work" (Hayles, 2017, p. 2), the author adheres to a definition of consciousness governed by a utilitarian performance-driven loop whereby our acts in / on the world become a (contingent) measure of our relevance therein. The relations entertained by entities and objects are hence made to depend on a discrete definition of identity as an atomic component of a "self-referential field" (Hayles, 2018, p. 10) where language is reduced to a code connecting a network of statements referring to themselves through a fuzzily defined continuity governed by family resemblance.

The question that arises is: How can Hayles' field metaphor account for the emergence and further deployment of eventful states to an Interpretant? For consciousness does not emerge as a result of a metalinguistic reconstruction of the process, but the definition of the meaningfulness of *its becoming* as part of an agent's (re)construction of identity; which, of necessity, is and will always be a measure of the mind defining the boundaries of the actions that guarantee its persistence in time and space. It is here that the notion of consciousness (and a definition thereof) becomes handy.

3. The uniqueness of human consciousness

Where, then, is consciousness to be found? What amounts to a conscious state, and what kind of consciousness can be said to provide the access of a system, biological or not, to a coupling of inner-outward experience? *According to Information Integration Theory* (*IIT 1.0*, Tononi, 2004; *ITT 3.0*, Oizumi et al., 2014; *ITT 4.0*, Albantakis et al. 2023), the question of consciousness needs to be defined from a first-person perspective. In this vein, IIT considers two problems: 1) the extent to which a system can be considered as having conscious experience, and 2) the definition of the conditions that determine the type of consciousness a system has. This phenomenological reading departs from the assumption that, for a system to be conscious of itself and of what is going on around it, it must, first, be able to deal with large amounts of information (differentiation), and second, be capable of integrating that information into a coherent semiotic picture of the world (integration). Moreover, the whole process is said to be private and occurs in specific areas of the brain.

Expectedly, the most significant weakness of IIT lies in its epistemological limitations, since it posits that systems such as photodiodes may possess consciousness, yet there is no independent method to verify such claims. Furthermore, the theory's central metric, Φ (integrated information), faces significant issues. Calculations of Φ often yield non-unique values that depend on arbitrary modeling choices, making it unclear how such results should be interpreted. Additionally, the range of Φ values lacks specificity, thereby preventing clear conclusions about the relationship between Φ and conscious experience. Philosophically, IIT introduces problematic and counterintuitive claims. For example, the theory implies that consciousness can be quantified, leading to assertions that a 2D grid might possess "10 times the amount of consciousness" as the human brain. Such statements are not only counterintuitive but also conceptually incoherent, as they fail to clarify what "amount of consciousness" means. Moreover, although, as it will become evident further below, I reject the idea of a Hard problem of consciousness (Chalmers 1996), IIT does not adequately address such problem (its stated goal), that is, the question of why and how physical processes give rise to subjective experiences. This leaves the theory vulnerable to the "zombie argument," wherein one can imagine a universe populated by high-Φ systems that exhibit no inner experience. These scenarios expose IIT's inability to account for the qualitative dimensions of consciousness, which remain essential to any comprehensive explanation.

It is no coincidence then that my selection of this specific theory of consciousness targets the consequences of reducing self-awareness to a series of symbolic operations subsumed to a computational bias mysteriously encoded in our brains, usually termed functional consciousness, that is, based on "mere functions and abilities" (see Hassel Mørch, 2023, p. 3).

The paradox here is that ITT sells itself as a phenomenal theory of consciousness by which "[p]henomenally conscious states are characterized by the fact that there is *something that it's like* for a creature or entity to be in them, or that they are subjectively experienced or felt" (Hassel Mørch 2023, 3; emphasis in original). And yet, on closer inspection, ITT's focus is primarily on the physical aspect of sentience, having the brain as a recipient of seeming subjectivity and self-awareness.

Thus, in a manner akin to the posthumanist musings reviewed previously, many studies have seen in ITT the opportunity to dissolve human singularity into the alleged materiality of a "self-conscious" wetware producing signal-driven informational interactions as a response to a stimulus (e.g., Montoya, 2023). Central to this claim is the idea that conscious responses are fired by a stimulus that can be further incorporated into the system to produce an experiential response. This leads us to a second objection to ITT: its reductionism of conscious states to an impoverished version of the brain as a physical support of experience (a quantity).

In contrast, the present proposal defines consciousness as a combination of four main elements: 1) The identification of objects in the world associated with events of perception (experience is intrinsic, though modeled on shared experiential structures in common with other organisms). 2) The construction of pictures of the world through the integration of modal and amodal information processes in the brain (information is provided by different modalities and then organized in the brain in specific areas. The nature of experience is defined by degrees of perceptual access to complexes (composite concepts) and not by simulations of experience in the form of "shadows of perception". 3) The content of experiential states becomes integrated into the agentic projection of intentional states onto the world. 4) The alignment of subjective conscious states with other conscious entities is facilitated by the construction of a model of the self both as individual and as part of a system of agentive relations and hierarchies. Agentive hierarchies are the instances by which decisions and non-decisions are made to coincide in order to reveal to an agent or an observer the purpose and reason for engaging in a particular action given a state of affairs within an event. As previously suggested, such delimitation requires a sense of boundary that separates, though not insulates, the feel of being in possession of a body that is distinct from the flux of energy surrounding it.

As we saw in *subsection 1.1.2*, the notion of bodily awareness as an instance of agency is partly indebted to the AIF theory (Friston, 2009, 2010), according to which organisms need to keep a balance between environmental and inner-body information in order to avoid entropy (systemic collapse). This defines the role of biological systems as predictive entities moving toward specific attracting states, that is, points of equilibrium that minimize entropy in order to attain homeostatic integrity. Homeostasis (inner-system equilibrium) is accomplished through the identification of sensory states by means of ad hoc receptors placed in a statistical boundary (Markov Blanket) providing a link between system-external and system-internal conditions.

To assess this in the context of human language production, we need to explain what perceptual experience is supposed to be. In the first place, language is constructed on concepts, and the existence of concepts point to a need to recognize the substance of objects and entities in the world. So, a definition of concepts is required at this point.

3.1. Defining Concepts:

In the context of intelligent agency, concepts can be defined as the sum of animate (natural) and inanimate (natural or constructed, aka. "extended") entities that provide agent-conceptualizers with perceptual maps for the reconstruction of reality. I classify concepts into three main categories: Prototypes, exemplars, and theories.

3.1.1. Prototypes

Prototypes have their origin in a basic set of idealized tendencies that are entrenched as the result of beliefs about intrinsic properties, that is, the *what* making up the substance of objects and entities in the world. These provide conceptualizers with sensory-driven, *part-to-whole* relations leading to semiotic integration. For example, bird, fish, or feline essences consist of particular features that are ascribed to a prototypical natural kind (NK):

Examples:

Feline Glance:

Essence_N: The idealized tendency associated with feline features (e.g., eye shape, expression).

Prototypical NK: The general concept of a cat, including shared characteristics.

Example: The way a cat looks at something, characterized by the specific shape and expression of its eyes, can be considered as a prototype that encapsulates the essence of feline visual characteristics.

Cat-like Gait:

Essence_N: The idealized tendency associated with the way a cat moves (e.g., posture, movement pattern).

Prototypical NK: The general concept of a cat's movement.

Example: The distinctive walking style of a cat, with its characteristic posture and movement, serves as a prototype representing the essence of feline locomotion.

3.1.2. Exemplars

Exemplars are extended sets of stored prototypes representing *whole-to-part* associations of both natural and inanimate kinds (concepts). In other words, an exemplar is the best sample of a set of prototypes. Exemplars integrate more formally somatosensory experience with higher order conceptualization.

3.1.3. Theories

Theories, or beliefs, represent the combination of perceptual mappings and experiences aimed at swiftly accessing categories. The assignment of categories involves a process of reality reconstruction, where the actions of the world upon natural kinds (including entities like humans) produce perceptually accessible effects. The essence of natural kinds resides at the intersection of cognition and patterns of biological adaptation to the environment, encompassing phylogenetic extension and ecomorphological convergence. Categories emerge from a synthesis of perception, beliefs, and our interactions with the world and other organisms. Natural kinds, being mind-independent, originate from our phenomenological grasp of ways of being in the world, facilitated by our sensitivity to various natural effects shaping the essence of organisms (confer Torres-Martínez, 2024c).

The identification of the essence of things goes beyond the simple categorization of observable features along a continuum of prototypical characteristics that entities possess. Indeed, unobservable essences are not homogeneous, linearly arranged properties of identifiable natural kinds, but rather fragments of experience retrievable across various layers of perceptual modeling in a bottom-up manner. The idea is that language connects these essences to observable surface traits through specific affordances, represented as a vector E= [e1, e2, ..., ek]. So the essential properties of a natural kind like a tiger are not just symbolic representations of intrinsic properties in a conceptualizer's mind but reflections of a network of potentialities inherent in events where these essences are active (Torres-Martínez 2024c).

It is important to see how much this objection militates against the possibility of postulating single-layered propositions for the description of intrinsic features in natural kinds. It is perhaps not surprising that defining the existence of an entity or object requires positing a multilayered set of propositions following a model like this: "If condition S becomes true upon the lower scale length ΔL, then condition S* will become true on the higher scale length ΔH" (Wilson, 2023, pp. xiv-xv).

On my account, then, a property essential at one scale does not prevent it from being intrinsic at another level (see Torres-Martínez, 2024b). For tigers, lacking stripes may not inherently negate core "tigerness," but could hamper camouflage and hunting success existentially. Therefore, the fur property cannot be separated from its functional value in the natural environment. As a result, statements like "Redness and squareness are intrinsic properties. Being next to a red object is extrinsic" (Vallentyne, 2014, p. 31) become nonsensical given the multilayered nature of linguistic reconstruction.

This reasoning has limits, since stripes and orange fur are intrinsic for successful hunting but potentially extrinsic from the perspective of prey like Chital that could be deceived by the camouflage. So it is inaccurate to say "[w]hether something is red, or 3 kg, or round is a matter of how it itself is, regardless of anything else" (Denby, 2014, p. 91).

4. Conclusion

Throughout this paper, I have challenged perspectives that attribute genuine intelligence, consciousness, or the capacity to construct meaning and culture to generative artificial intelligence, particularly large language models. This becomes clearer when considering the epistemic blindness pervading the dehumanizing tenor of current linguistic and neuroscientific research, which can only be interpreted as a strategy to amplify these fields' influence on other epistemic communities through the imposition of their own methods, ontological formulations, and agendas (e.g. Tuckute, 2024). This is compounded by narratives that obscure the human experience framed within a posthumanist discourse that erroneously posits machinic information integration and performance as a prerequisite for sentience and agency. This has been interpreted as a threat to science in many respects, requiring analysis from philosophers of science who warn against the proliferation of *science-before-testimony* that reduces the Human to data and perplexity values (see Torres-Martínez, 2024c). I have argued that a means to correct this trend is to place the focus on the definition of foundational concepts such as consciousness, natural kinds, and concept formation. To underscore the central argument, the thesis has been that human consciousness and agency are intrinsically tied to our biological architecture through the holistic integration of perception, conceptualization, intentional action, and self-awareness.

It is instructive to remember that, in the absence of a clear distinction between the human and the machinic, it becomes painfully easy to fall into the conceptual trap of unfounded sentience attribution. Evidently, advocates of information integration as a purported sign of consciousness quickly run out of defensible ground, since, although computational models may simulate aspects of language processing, they fundamentally lack the grounded, embodied experience and drive that shape human linguistic behavior. This lack of alignment, that is, the requirement on the part of (human) *choice architects* (see Mills & Skaug Sætra, 2024) to align with human values and preferences, departs from the premise that "it is logically necessary that AI does not itself substantially shape the very cognitive resources with which it aims to match; if it does, alignment, for example, would become recursive and incoherent, i.e., AI aligning with itself." (Wihbey, 2024, p. 4).

AI-self-alignment, as supported by some cognitive linguists and neuroscientists poses a threat to the persistence of human knowledge construction and the values providing meaningfulness to the process. This advances us toward an emerging realization that language arises from an organism's need to reduce uncertainty about its environment and sustain itself through accurate predictive modeling that reflects the speakers' worldview and agentivity. Alternatively stated, linguistic constructions emerge from an embodied, agent-driven process rather than purely statistical operations over data (contra Goldberg, 2019). This rejects both the empiricist notion that language and meaning can be entirely captured by information processing machines, as well as posthumanist views that conflate sign production with conscious meaning-making. Thus, we may take it as substantiated that the crux here is that sign systems manifest purposeful action but are not equivalent to action itself, which contends the idea that "the post-humanist movement aims to question human exceptionalism and the role that modernity has given to human beings." (Gómez Redondo et al. 2024, p. 12). This interpretation has considerable merit, for example, to reveal that the attribution of agency and genuine semiosis to artificial systems is a conceptual overreach stemming from folk, before-testimony theorizing that fails to grasp the nuances of biological consciousness.

Ultimately, this paper has characterized *the Human* as a manifold of qualities and properties associated with an intrinsically embodied species whose linguistic capacities are profoundly intertwined with their biological identities as conscious, sense-making agents tuned to the world. As has been remarked before, the extended humanist position embraced herein stands in stark contrast to reductive perspectives championing the deconstruction of "binarism" and "normalized selves" as a means to impose a blind neo-*triadism* wherein the "Other" (the permanently undefined *Third*, including machines) is solely construed in terms of its capacity to introduce ambiguity and disruption, while failing to recognize other subjectivities as inherent rather than obstructive to a strategy "to disrupt a humanist version of being" (Toffoletti, 2007, p. 84). While generative AI can simulate surface behaviors and smooth the introduction of "a model of the self as either entirely resisting or complying with particular aspects of culture" (Toffoletti, 2007, p. 91), it cannot replicate the holistic experiences and organic self-modeling that fundamentally underlie human language and meaning construction, irrespective of how the Human is conceptualized.

References

Albantakis, L., Barbosa, L., Findlay, G., Grasso, M., Haun, A. M., Marshall, W., et al. (2023). Integrated information theory (IIT) 4.0: Formulating the properties of phenomenal existence in physical terms. *PLoS Computational Biology, 19*(10), Article e1011465. https://doi.org/10.1371/journal.pcbi.1011465

Andersson, P., Strandman, A., & Strannegård, C. (2019). Exploration strategies for homeostatic agents. In *Artificial General Intelligence: 12th International Conference* (pp. 178-187).

Arnold, T. (2022). The tragedy of scientific culture: Husserl on inauthentic habits, technisation and mechanisation. *Human Studies, 45*, 209–222. https://doi.org/10.1007/s10746-022-09621-x

Balduin-Philipps, L. S., Weiss, S., Schaller, F., & Müller, H. M. (2011). Abstract action language processing in eleven-year-old children: Influence of upper limb movement on sentence comprehension. *Behavioral Sciences, 11*(12), Article 162. https://doi.org/10.3390/bs11120162

Barsalou, L. W. (1982). Context-independent and context-dependent information in concepts. *Memory & Cognition, 10*, 82–93. https://doi.org/10.3758/BF03197629

Beuls, K., & Van Eecke, P. (2024). Humans learn language from situated communicative interactions: What about machines? *Computational Linguistics, 50*(4), 1-35. https://doi.org/10.1162/coli_a_00534

Beuls, K., & Van Eecke, P. (2025). Construction grammar and artificial intelligence. In M. Fried & K. Nikiforidou (Eds.), *Cambridge handbook of construction grammar*. [Preprint]. arXiv:2309.00135

Block, N. (1995). On a confusion about the function of consciousness. *Behavioral and Brain Sciences, 18*, 227–247. https://doi.org/10.1017/S0140525X00038188

Boiko, D. A., MacKnight, R., Kline, B., & Gomes, G. (2023). Autonomous chemical research with large language models. *Nature, 624*, 570-582. https://doi.org/10.1038/s41586-023-06792-0

Braidotti, R. (2013). *The posthuman*. Polity Press.

Buckner, C. J. (2023). *From deep learning to rational machines: What the history of philosophy can teach us about the future of artificial intelligence*. Oxford University Press.

Butlin, P. (2022). Machine learning, functions and goals. *Croatian Journal of Philosophy, 22*(66), 351–370. https://doi.org/10.52685/cjp.22.66.5

Catak, F. O., & Kuzlu, M. (2024). Uncertainty quantification in large language models through convex hull analysis. *Discover Artificial Intelligence,* 4(90), 1-14. https://doi.org/10.1007/s44163-024-00200-w

Cave, S., & Dihal, K. (Eds.). (2023). *How the world sees intelligent machines: Introduction. In Imagining AI: How the world sees intelligent machines* (pp. 3-15). Oxford University Press.

Chalmers, D. J. (1996). *The conscious mind: In search of a fundamental theory*. Oxford University Press.

Chalmers, D. J. (2014). *Constructing the world*. Oxford University Press.

Chomsky, N., Gallego, Á. J., & Ott, D. (2019). Generative grammar and the faculty of language: Insights, questions, and challenges. *Catalan Journal of Linguistics, Special Issue*, 229-261. https://doi.org/10.3390/e25091328

Clarke, B. (2008). *Posthuman metamorphosis: Narrative and systems*. Fordham University Press.

Clusmann, J., Kolbinger, F. R., Muti, H. S., Carrero, Z. I., Eckardt, J-N., Laleh, N. J., Löffler, C. M. L., Schwarzkopf, S. C., Unger, M., Veldhuizen, G. P., Wagner, S. J., & Kather, J. N. (2023). The future landscape of large language models in medicine. *Communications Medicine, 3*(141), 1-8. https://doi.org/10.1038/s43856-023-00370-1

Coole, D., & Frost, S. (Eds.). (2010). Introducing new

materialisms. In *New materialism: Ontology, agency and politics* (pp. 1-43). Duke University Press.

Cortez, N., & Sage, W. M. (2023). The disembodied First Amendment. *Washington University Law Review, 707*. https://ssrn.com/abstract=4406190

Curto, G., Jojoa Acosta, M. F., Comim, F., & Garcia-Zapirain, B. (2024). Are AI systems biased against the poor? A machine learning analysis using Word2Vec and GloVe embeddings. *AI & Society, 39*, 617–632. https://doi.org/10.1007/s00146-022-01494-z

Dehaene, S., Lau, H., & Kouider, S. (2017). What is consciousness, and could machines have it? *Science, 358*, 486–492. https://doi.org/10.1126/science.aan8871

Demuro, E., & Gurney, L. (2023). Can nonhumans speak? Languaging and worlds in posthumanist applied linguistics. *Linguistic Frontiers, 6*(2), 94-105. https://doi.org/10.1007/s11229-024-04566-3

Demuro, E., & Gurney, L. (2024). Artificial intelligence and the ethnographic encounter: Transhuman language ontologies, or what it means "to write like a human, think like a machine." *Language & Communication, 96*, 1-12. https://doi.org/10.1146/annurev-neuro-120623-101142

Desmond, H., & Ramsey, G. (Eds.). (2023). Introduction: The manifold challenges to understanding human success. In *Human success: Evolutionary origins and ethical implications* (pp. 1-14). Oxford University Press.

Dulberg, Z., Dubey, R., Berwian, I. M., & Cohen, J. D. (2023). Having multiple selves helps learning agents explore and adapt in complex changing worlds. *Proceedings of the National Academy of Sciences, 120*(28), Article e2221180120.

Duncker, D. (2020). Chatting with chatbots: Sign making in text-based human-computer interaction. *Sign Systems Studies, 48*(1), 79-100. https://doi.org/10.12697/SSS.2020.48.1.05

Dynel, M. (2023). Lessons in linguistics with ChatGPT: Metapragmatics, metacommunication, metadiscourse and metalanguage in human-AI interactions. *Language & Communication, 93*, 107–124. https://doi.org/10.1007/s001090000086

Fedorenko, E., Ivanova, A., & Regev, T. (2024). The language network as a natural kind within the broader landscape of the human brain. *Nature Reviews Neuroscience, 5*, 289–312. https://doi.org/10.1038/s41583-024-00802-4

Friedrich, J., Golkar, S., Farashahi, S., Genkin, A., Sengupta, A., & Chklovskii, D. (2021). Neural optimal feedback control with local learning rules. *Advances in Neural Information Processing Systems, 34*, 16358–16370. https://dl.acm.org/doi/10.5555/3540261.3541512

Friston, K. (2009). The free-energy principle: A rough guide to the brain? *Trends in Cognitive Science, 13*, 293–301. https://doi.org/10.1016/j.tics.2009.04.005

Friston, K. (2010). The Free-Energy Principle: A unified brain theory? *Nature Review of Neuroscience, 11*, 127–138. https://doi.org/10.1038/nrn2787

Gerken, M. (2022). *Scientific realism: Its roles in science and society*. Oxford University Press.

Glenberg, A. M., & Kaschak, M. P. (2002). Grounding language in action. *Psychonomic Bulletin & Review, 9*, 558–565. https://doi.org/10.3758/BF03196313

Goldberg, A. E. (1995). *Constructions: A construction grammar approach to argument structure*. University of Chicago Press.

Goldberg, A. E. (2019). *Explain me this: Creativity, competition, and the partial productivity of constructions*. Princeton University Press.

Gómez Redondo, S., Rodríguez Higuera, C. J., Coca, C. R., & Olteanu, A. (2024). Transhumanism, society and education: An edusemiotic approach. *Studies in Philosophy and Education*. https://doi.org/10.1007/s11217-024-09927-6

Harding, M. E. (2020). *The 'I' and the 'Not-I': A study in the development of consciousness (Bollingen Series LXXIX)*. Princeton University Press. (Original work published 1965)

Hayles, N. K. (1999). *How we became posthuman: Virtual bodies in cybernetics, literature, and informatics*. University of Chicago Press.

Hayles, N. K. (2012). *How we think: Digital media and contemporary technogenesis*. University of Chicago Press.

Hayles, N. K. (2017). *Unthought: The power of the cognitive nonconscious*. University of Chicago Press.

Hayles, N. K. (2018). *The cosmic web: Scientific field models and literary strategies in the twentieth century*. Cornell University Press.

Hayles, N. K. (2021). *Postprint: Books and becoming computational*. Columbia University Press.

Hayles, N. K. (2024). Posthuman bodies: Why they (still) matter. In G. Hamilton & C. Lau (Eds.), *Mapping the posthuman* (pp. 29-48). Routledge.

Hassel Mørch, H. (2023). *Non-physicalist theories of consciousness*. Cambridge University Press.

He, C., Chrastil, E. R., & Hegarty, M. (2022). A new psychometric task measuring spatial perspective taking in ambulatory virtual reality. *Frontiers in Virtual Reality, 3*, Article 971502. https://doi.org/10.3389/frvir.2022.971502

Hedlund, M., & Persson, E. (2024). Expert responsibility in AI development. *AI & Society, 39*, 453-464. https://doi.org/10.1007/s00146-022-01498-9

Jablonka, K. M., Schwaller, P., Ortega-Guerrero, A., & Smit, B. (2023). Leveraging large language models for predictive chemistry. *Nature Machine Intelligence, 6*, 161-169. https://doi.org/10.1038/s42256-023-00788-1

Jackendoff, R. (2012). *A user's guide to thought and meaning*. Oxford University Press.

Juechems, K., & Summerfield, C. (2019). Where does value come from? *Trends in Cognitive Sciences, 23*(10), 836–850. https://doi.org/10.1016/j.tics.2019.07.012

Kean, H., Fung, A., Pramod, R. T., Chomik-Morales, J., Kanwisher, N., & Fedorenko, E. (2024). Intuitive physical reasoning is not mediated by linguistic nor exclusively domain-general abstract representations. https://doi.org/10.1101/2024.11.25.625212

Keramati, M., & Gutkin, B. (2014). Homeostatic reinforcement learning for integrating reward collection and physiological stability. *eLife, 3*, Article e04811. https://doi.org/10.7554/eLife.04811

Kessler, K., & Thompson, L. A. (2014). The embodied nature of spatial perspective taking: Embodied transformation versus sensorimotor interference. *Cognition, 114*, 72–88. https://doi.org/10.1016/j.cognition.2009.08.015

Kinderman, D., & Onofri, A. (2021). The fragmented mind: An introduction. In C. Borgoni, D. Kindermann, & A. Onofri (Eds.), *The fragmented mind* (pp. 1-33). Oxford University Press.

Kinouchi, Y., & Mackin, K. J. (2018). A basic architecture of an autonomous adaptive system with conscious-like function for a humanoid robot. *Frontiers in Robotics, 5*, Article 30. https://doi.org/10.3389/frobt.2018.00030

Kitano, H. (2021). Nobel Turing Challenge: Creating the engine for scientific discovery. *npj Systems Biology and Applications, 7*(29), 1-12. https://doi.org/10.1038/s41540-021-00189-3

Kumar, P. (2024). Large language models (LLMs): Survey, technical frameworks, and future challenges. *Artificial Intelligence Review, 57*(260), 1-51. https://doi.org/10.1007/s10462-024-10888-y

Levy, D. (2009). The ethical treatment of artificially

conscious robots. *International Journal of Social Robotics, 1*, 209–216. https://doi.org/10.1007/s12369-009-0022-6

Luo, X., Rechardt, A., Sun, G., Nejad, K. K., Yáñez, F., Yilmaz, B., Lee, K., Cohen, A. O., Borghesani, V., Pashkov, A., Marinazzo, D., Nicholas, J., Salatiello, A., Sucholutsky, I., Minervini, P., Razavi, S., Rocca, R., Yusifov, E., Okalova, T., ... Love, B. C. (2024). Large language models surpass human experts in predicting neuroscience results. *Nature Human Behaviour.* https://doi.org/10.1038/s41562-024-02046-9

Lux, V., Non, A. L., Pexman, P. M., Stadler, W., Weber, L. A. E., & Krüger, M. (2021). A developmental framework for embodiment research: The next step toward integrating concepts and methods. *Frontiers in Systems Neuroscience, 15*, Article 672740. https://doi.org/10.3389/fnsys.2021.672740

Machery, E. (2010). Reply to Barbara Malt and Jesse Prinz. *Mind & Language, 25*(5), 634-646.

Mahon, B. Z. (2008). Action recognition: Is it a motor process? *Current Biology, 18*(22), R1068-R1069. https://doi.org/10.1016/j.cub.2008.10.00

Mahon, B. Z. (2015a). The burden of embodied cognition. *Canadian Journal of Experimental Psychology, 69*(2), 172–178. https://doi.org/10.1037/cep0000060

Mahon, B. Z. (2015b). Response to Glenberg: Conceptual content does not constrain the representational format of concepts. *Canadian Journal of Experimental Psychology, 69*(2), 179–180. https://doi.org/10.1037/cep0000059

Mahon, B. Z. (2015c). What is embodied about cognition? *Language, Cognition and Neuroscience, 30*(4), 420–429. https://doi.org/10.1080/23273798.2014.987791

Mahon, B. Z., & Hickok, G. (2016). Arguments about the nature of concepts: Symbols, embodiment, and beyond. *Psychonomic Bulletin Review, 23*, 941-958. https://doi.org/10.3758/s13423-016-1045-2

Massimi, M. (2022). *Perspectival realism*. Oxford University Press.

McDowell, J. (1992). Meaning and intentionality in Wittgenstein's later philosophy. In P. French, T. Uehling, & H. Wettstein (Eds.), *Midwest studies in philosophy* (Vol. 17, pp. 40-52). University of Notre Dame Press.

Mills, S., & Skaug Sætra, H. (2024). The autonomous choice architect. *AI & Society, 39*, 583–595. https://doi.org/10.1007/s00146-022-01486-z

Montoya, I. (2023). What is it like to be a brain organoid? Phenomenal consciousness in a biological neural network. *Entropy, 25*, Article 1328. https://doi.org/10.3390/e25091328

Mougenot, D., & Matheson, H. (2024). Theoretical strategies for an embodied cognitive neuroscience: Mechanistic explanations of brain-body-environment systems. *Cognitive Neuroscience*. https://doi.org/10.1080/17588928.2024.2349546

Murphy, E., Holmes, E., & Friston, K. (2024). Natural language syntax complies with the free energy principle. *Synthese, 203*(154), 1-35. https://doi.org/10.1007/s11229-024-04566-3

Nov, O., Singh, N., & Mann, D. (2023). Putting ChatGPT's medical advice to the (Turing) test. *arXiv*. https://doi.org/10.48550/arXiv.2301.10035

Oizumi, M., Albantakis, L., & Tononi, G. (2014). From the phenomenology to the mechanisms of consciousness: Integrated Information Theory 3.0. *PLoS Computational Biology, 10*(5), Article e1003588. https://doi.org/10.1371/journal.pcbi.1003588

Pan, X., Liang, B., & Li, X. (2024). Flexible and fine-grained simulation of speed in language processing. *Frontiers in Psychology, 15*, Article 1333598. https://doi.org/10.3389/fpsyg.2024.1333598

Reber, A. S., Baluška, F., & Miller Jr., W. B. (2023). *The sentient cell: The cellular foundations of consciousness*. Oxford University Press.

Rickabaugh, B., & Moreland, J. P. (2023). *The substance of consciousness: A comprehensive defense of contemporary substance dualism*. Wiley-Blackwell.

Rosendahl Thomsen, M. (2013). *The new human in literature: Posthuman visions of changes in body, mind and society after 1900*. Bloomsbury Academic.

Searle, J. R. (1980). Minds, brains and programs. *Behavioral and Brain Sciences, 3*, 417–424. https://doi.org/10.1017/S0140525X00005756

Searle, J. R. (1992). *The rediscovery of the mind*. MIT Press.

Schmidgall, S., Harris, C., Essien, I., Olshvang, D., Rahman, T., Kim, J. W., Ziaei, R., Eshraghian, J., Abadir, P., & Chellappa, R. (2024). Evaluation and mitigation of cognitive biases in medical language models. *npj Digital Medicine, 7*(295), 1-9. https://doi.

org/10.1038/s41746-024-01283-6

Shrader-Frechette, K. (2014). *Tainted: How philosophy of science can expose bad science*. Oxford University Press.

Sims, M., & Kiverstein, J. (2021). Externalized memory in slime mould and the extended (nonneuronal) mind. *Cognitive Systems Research, 73*, 26-35. https://doi.org/10.1016/j.cogsys.2021.12.001

Solaki, A. (2022). The effort of reasoning: Modelling the inference steps of boundedly rational agents. *Journal of Logic, Language and Information, 31*, 529–553. https://doi.org/10.1007/s10849-022-09367-w

Speed, L. J., & Vigliocco, G. (2014). Eye movements reveal the dynamic simulation of speed in language. *Cognitive Science, 38*, 367–382. https://doi.org/10.1111/cogs.12096

Todorov, E., & Jordan, M. I. (2002). Optimal feedback control as a theory of motor coordination. *Nature Neuroscience, 5*(11), 1226-1235. https://doi.org/10.1038/nn963

Toffoletti, K. (2007). *Cyborgs and Barbie dolls: Feminism, popular culture and the posthuman body*. I.B. Tauris.

Tononi, G. (2004). An information integration theory of consciousness. *BMC Neuroscience, 5*, Article 42. http://www.biomedcentral.com/1471-2202/5/42

Torres-Martínez, S. (2018a). Constructions as triads of form, function and agency: An agentive cognitive construction grammar analysis of English modals. *Cognitive Semantics, 4*(1), 1–38. https://doi.org/10.1163/23526416-00401001

Torres-Martínez, S. (2018b). Exploring attachment patterns between multi-word verbs and argument structure constructions. *Lingua, 209*, 21–43. https://doi.org/10.1016/j.lingua.2018.04.001

Torres-Martínez, S. (2019). Taming English modals: How a construction grammar approach helps to understand modal verbs. *English Today, 35*(2), 50–57. https://doi.org/10.1017/S0266078418000081

Torres-Martínez, S. (2020). On English modals, embodiment and argument structure: Response to Fong. *English Today, 38*(2), 105–113. https://doi.org/10.1017/S0266078420000437

Torres-Martínez, S. (2021a). The cognition of caused-motion events in Spanish and German: An agentive cognitive construction grammar analysis. *Australian Journal of Linguistics, 41*(1), 33–65. https://doi.org/10.1080/07268602.2021.1888279

Torres-Martínez, S. (2021b). Complexes, rule-following, and language games: Wittgenstein's philosophical method and its relevance to semiotics. *Semiotica, 242*, 63–100. https://doi.org/10.1515/sem-2019-0113

Torres-Martínez, S. (2022a). Metaphors are embodied otherwise they would not be metaphors. *Linguistics Vanguard, 8*(1), 185–196. https://doi.org/10.1515/lingvan-2019-0083

Torres-Martínez, S. (2022b). The role of semiotics in the unification of Langue and Parole: An agentive cognitive construction grammar approach to English modals. *Semiotica, 244*(1/4), 195–225. https://doi.org/10.1515/sem-2018-0046

Torres-Martínez, S. (2022c). On the cognitive dimension of metaphors and their role in education: A response to Molina Rodelo (2021). *Revista Senderos Pedagógicos, 13*, 113–123. https://doi.org/10.53995/rsp.v13i13.1128

Torres-Martínez, S. (2023a). A radical embodied characterization of German Modals. *Cognitive Semantics, 9*(1), 132–168. https://doi.org/10.1163/23526416-bja10035

Torres-Martínez, S. (2023b). The semiotics of motion encoding in early English: A cognitive semiotic analysis of phrasal verbs in old and middle English. *Semiotica, 251*, 55–91. https://doi.org/10.1515/sem-2019-0104

Torres-Martínez, S. (2023c). Grammaire agentielle cognitive de constructions: Explorations sémioticolinguistiques des origines de la représentation incarnée. *Signata, Annales de Sémiotique, 14*. https://doi.org/10.4000/signata.4551

Torres-Martínez, S. (2024a). Embodied human language models vs. large language models, or why Artificial Intelligence cannot explain the modal *be able to*. *Biosemiotics, 17*, 185–209. https://doi.org/10.1007/s12304-024-09553-2

Torres-Martínez, S. (2024b). Embodied essentialism in the reconstruction of the animal sign in robot animal design. *Biosystems, 238*, Article 105178. https://doi.org/10.1016/j.biosystems.2024.105178

Torres-Martínez, S. (2024c). Semiosic translation: A Bayesian-heuristic theory of translation and translating. *Language and Semiotic Studies, 10*(2), 167–202. https://doi.org/10.1515/lass-2023-0042

Torres-Martínez, S. (2024d). A predictive human model of language challenges traditional views in linguistics and pretrained transformer research. *Language and Semiotic Studies*. https://doi.org/10.1515/lass-2024-0018

Tuckute, G. [@GretaTuckute]. (2024, May 13). *Interested in supporting research at the intersection of cognitive science, neuroscience*, and AI? [Tweet]. X.

Tuckute, G., Kanwisher, N., & Fedorenko, E. (2024). Language in brains, minds and machines. *Annual Review of Neuroscience, 47*. https://doi.org/10.1146/annurev-neuro-120623-101142

Turkle, S. (2005). *The second self: Computers and the human spirit*. MIT Press.

Turkle, S. (2010). *Alone together: Why we expect more from technology and less from each other*. Basic Books.

Vannuscorps, G., Rombaux, E., Andres, M., Pereira Carneiro, S., & Caramazza, A. (2021). Typically efficient lipreading without motor simulation. *Journal of Cognitive Neuroscience, 33*(4), 611–621. https://doi.org/10.1162/jocn_a_01666

Vickers, P. (2023). *Identifying future-proof science*. Oxford University Press.

Weissweiler, L., Köksal, A., & Schütze, H. (2023). Hybrid Human-LLM corpus construction and LLM evaluation for rare linguistic phenomena. *arXiv*:2403.06965

West, O. (2024). What makes a good theory, and how do we make a theory good? *Computational Brain and Behavior*. https://doi.org/10.1007/s42113-023-

00193-2

Wihbey, J. P. (2024). AI and epistemic risk for democracy: A coming crisis of public knowledge? *Ethics Institute Working Paper*. https://ssrn.com/abstract=4805026

Wilson, M. (2023). *Imitation of rigor: An alternative history of analytic philosophy*. Oxford University Press.

Wittgenstein, L. (1984). *Werkausgabe Band 1. Tractatus Logico-Philosophicus/Tagebücher/Philosophische Untersuchungen. Suhrkamp Verlag*. (Original work published 1953)

Youssef, A., Stein, S., Clapp, J., & Magnus, D. (2023). The importance of understanding language in large language models. *The American Journal of Bioethics, 23*(10), 6-7. https://doi.org/10.1080/15265161.2023.2256614

CHAPTER 4

Credence, attribution and creativity in the construction of literary meaning with *generative artificial intelligence*

Abstract

The rise of *generative artificial intelligence* (GenAI) has shifted the perception of machines from mere tools to agents capable of constructing culture, with disembodied language models (LLMs) increasingly viewed as reliable mirrors of the human mind. This reduction of the mind to a computational model raises questions about the role of human consciousness, experience, and agency. I argue that, by embracing a posthuman computational view, we risk to dissolve humanness into algorithmic processes devoid of phenomenological depth, thereby grounding meaning in abstract symbolic representations rather than embodied, metaphorical reasoning tied to physical and social realities. This shift towards an algorithmic mindset, driven by the promise of efficiency and performance, also risks exhausting the narratological credibility of plots introducing possible worlds constructed through technology. Stories once anchored in shared cultural metaphors could become mere algorithmic pastiches – fragmented collages devoid of meaning. This conceptual study explores how attributing agency to cognitively enhancing technologies results in an undue attribution of qualia to GenAI. It also examines reductive narratives of AI creativity based solely on information processing, contrasting this with human creativity based on divergent heuristics and reasoning through associative learning. Ultimately, while AI can enhance human capabilities, over-anthropomorphizing these systems risks ascribing qualities these symbolic models lack, which requires us to devise a clear delineation of boundaries to harness GenAI's power responsibly.

Keywords: AI agency attribution, AI-assisted literature, AI-tool credence, generative AI, large language models, planned incongruity through model hallucination,

1 Introduction

The advent of *generative Artificial Intelligence* (GenAI) has moved the dial of credence in the power of the *machinic-as-tool* to the attribution of sentience to the *machine-as-agent*. The ensuing AI sensationalism or "AI hype" (Vrabič Dežman 2024) has led to a perception that human-machine interaction as a true interaction between deep learning technologies and humans. Seen through this prism, AI technologies are defined as agents capable of constructing culture in the context of posthuman (antihumanist) narratives aiming to remove human exceptionalism and uniqueness for the sake of a revisionist dissolution of the social subject into logocentric "practice, performance, [and] enactment" (Demuro and Gurney 2023: 96). This emerging "liberal humanist subject with its own autonomous substantiality (essentiality) and the capacity for self-governance and as a rational and integral being with clearly defined boundaries" (Stasieńko 2022: 2) becomes a relational subject (Braidotti 2013: 26) deprived of symbolic contours and significance beyond the construction of the human through an undefined "*other*" (the non-human comprising animals and machines). Thus, "[f]or posthuman theory, the subject is a transversal entity, fully immersed in and immanent to a network of non-human (animal, vegetable, viral) relations (Braidotti, 2013, p. 193).

Among other things, this essentialist ideology has provided the impetus for a growing identification of disembodied models of language generation, such as large language models (henceforth LLMs, e.g. OpenAI's GPT series, Google's Gemini Ultra, Anthropic's Claude 3 Opus, etc.), as mirrors of the human mind conceived of as a non-exceptional mathematical decision-making vector space for propositions and logical reasoning.

LLMs such as ChatGPT are extensive neural networks exhibiting a large parametric complexity, often encompassing up to a trillion interconnected nodes. These architectures are expressly engineered for text processing and generation at high speed and context-bound accuracy. The seeming "eventful" performance of these models has derived in a perception of LLMs as intelligent agents capable of defining specific cultural trends and practices. The ensuing reduction of the human mind to a purely computational model through folk theorizing, that is, "an inarticulate yet influential *science-before-testimony* picture" (Gerken 2022: 1; emphasis in original) raises questions about the nature of consciousness, experience, and agency and the role of humans in a post-Anthropocene world a "*post*human time, a world without and after humans" (Colebrook and Weinstein 2017: x).

As we will see further below, by accepting the computational view, we are forced to embrace the further dissolution of humanness into a series of algorithmic processes devoid of phenomenological depth. Rather than grounding our understanding of the world in embodied, metaphorical reasoning tied to our physical and social realities, we find ourselves increasingly relying on abstract symbolic representations generated by artificial systems, which, quite paradoxically, are still far from mastering logical reasoning (Chen & Gao, 2024, p. 64). This shift towards an algorithmic mindset, propelled by the functionalist promise of greater efficiency and performance, also risks exhausting the narratological accountability of the possible worlds we construct through technology. Hence, our stories, once anchored in shared cultural metaphors and moral frameworks, could eventually become mere algorithmic collages of statistically probable outcomes devoid of deeper meaning and purpose (averageness)[1]. In this context, we must carefully consider the implications of ceding too much ground to GenAI systems in defining the contours of human experience and the symbolic foundations of human cultures in literary texts.

In what follows, I shall interrogate the emerging symbolic power of GenAI in the construction of literary

1 The integration of artificial intelligence into human cognitive processes poses a significant epistemological risk: the standardization of outputs toward mediocrity. This phenomenon is particularly evident in medicine, where studies have demonstrated AI's tendency to produce averaged, flattened representations of complex realities (Clusman et al., 2023; Schmidgall et al., 2024). This *convergence toward the mean* threatens to become the default paradigm across all domains of human endeavor, including scientific inquiry. Most concerning are the assertions from AI proponents who explicitly advocate for the subordination of human cognition to artificial intelligence systems. These advocates envision AI as the primary driver of scientific breakthrough by positioning it as a solution to human cognitive limitations through what Kitano (2021, p. 1) terms a "Turing grand challenge" - specifically, "a grand challenge to develop AI systems that can make significant scientific discoveries that can outperform the best human scientist."

texts. I argue that while traditional technologies revolved around a semiotic translation of text-based sign systems into various analogical and digital grammars, GenAI tools redefine text as a disembodied, logocentric, and unconstrained Escherian collection of events deployed within the confines of prompts. This redefinition is said to introduce a blurring of the line between organic conscious agents and artificial semiotic simulators. With this in mind, my proposal develops through several stages. In section two, I explore the various ways our penchant for attributing credence and agency to cognitively enhancing technologies operates an undue blending of purpose and tool affordance, thereby resulting in attribution of qualia (intrinsic qualities) to GenAI technologies (section three). In section four, I examine AI narratives associated with the alleged superiority of AI creativity, ushered in by a reductive reading of creative behavior as output-oriented performance based solely on information processing and integration. Finally, in section five, I analyze the role of creativity, with special emphasis on divergent reasoning as a hallmark of *the Human*. I argue that claims to "AI creativity" voiced by some researchers miss the point of creative behavior by conflating *AI hallucination* with the associative learning skills possessed by humans. The paper concludes with a reflection on the nature of AI as a tool for agentive enhancement that needs to be handled with restricted credence and attribution. While AI can augment and extend human capabilities, we must be aware of the risks of over-anthropomorphizing these systems and ascribing them qualities they do not truly possess. By delineating the boundaries between human cognition and artificial intelligence, we can unleash the power of GenAI tools while retaining a clear understanding of their limitations and the unique strengths of human intelligence, such as contextual awareness, associative reasoning, and the intentional exploration of incongruities – elements that are crucial for genuine creativity and innovation.

2 Credence

It is a deep fact about human experience that our representation of the world is shaped by different degrees of credence. One specific set of human credence pertains to the tools we use to extend our agency for task optimization and outcome. However, heuristic inference, that is, clue-driven decision-making, operates independently of a bias toward mathematical simplicity—mechanized, meaningless habits confined to closed physical systems. This is because pattern recognition in complex biological systems relies on bottom-up processes, identifying elements from environmental signals that facilitate the emergence of minimal working systems. These systems, conceptual frameworks enabling agents to experience consciousness-as-action, exemplify the adaptive nature of biological inference.

In the realm of creativity, this suggests that AI-generated outputs offer minimal working systems capable of supporting clue-based inference without necessitating the reduction of world complexity. Consequently, the interplay between creativity perception, content generation, and machinic representation has come to be understood as a graded dimension of attribution—that is, the assignment of properties to tools. This perspective has led some researchers to conclude, for instance, that "people prefer AI-generated poetry to human-authored poetry, consistently rating AI-generated poems more highly than the poems of well-known poets across a variety of qualitative factors" (Porter & Machery, 2024, p. 9).

However, the preference for AI-generated content among chatbot users does not reflect poor heuristics but rather a diminished investment in contextual analysis. This occurs through an enhanced attribution of precision and accuracy to AI technologies, which users perceive as tools for precision optimization and human outperformance (Jablonka et al., 2023). This phenomenon aligns with Arnold's critique of "empty signs," where habitual use of signs leads to an absence of meaningful engagement: "By using signs emptily, we do not even terminate in the consciousness of whatever it signifies anymore, because we use the habitual sign itself habitually, we do not invest mental activity into bringing the signified into focus, since we 'know' what it means" (Arnold, 2022, p. 216–217).

On this construal, our need to reduce the burden of investing too much cognitive capital in our "quest for the truth," results in inaccurate definitions of intelligence reduced to the very means we employ to amass symbolic capital and power over our environment and peers. The evolution of language and its subsequent institutionalization as a norm-making code through technologies such as the book format has thus far mirrored a common base of experience in the form of, for example, a range of metaphors associated with our intuitive physics, including modalities such as containment, path, trajectory, balance, etc. Thus, in the analogic world, our selected symbolic code for sharing, language, has served the role of a system for attunement with the world from which our experience-related concepts and consciousness are extracted and ultimately shaped.

Yet, the advent of the digital has brought about a redefinition of these modalities as a means to match the affordances of devices providing texts with a support that is, in turn, regulated by specific instructions (algorithms) aimed at providing an efficient rendering of the events being presented to the a reader. This comes with a transformation of the Interpretant-reader who is prompted to adjust their perceptual apparatus and cognitive abilities to the sort of simulations provided by the machine. The result is an increasing demand for credence from the Interpretant (the conceptualizing sign in the Peircean semiotic taxonomy) that, nonetheless, reduces the spectrum of possibilities for conceptualizers regarding the contours and scope of a character's agency and the necessary definition of the structure of events and the nature of specific states of consciousness related to a narratological baseline.

It is useful to highlight that the characterization of agents in contemporary movies and books stems from an increasingly restricted set of modalities such as force, path, trajectory (linked with action upon the world), and only rarely on balance and containment (associated with awareness and self-definition). It might thus be argued that this solipsistic redefinition of the human in a posthuman context has thus required the dissolution of humanness into relevant stages of calculated action leading to reduced cognitive effort within a sequence of events (instead of states) for performance and outcome optimization.

Most dramatically, the advent of GenAI has pushed the boundaries of credence to the realm of *attribution*. Another way of putting it is to say that human-machine interaction is now being defined by the identification of, for example, chatbots, as agents capable of constructing culture, or the possibility of disembodied models of language of defining the human mind as a mathematical decision-making cloud of propositions. This becomes clearer if one brings into discussion the notion of "engineering empiricism" stemming from an oversimplified view of the relationship between the success of deep learning systems and how humans and animals actually solve problems (Buckner, 2023, p. 3-4). As a result, the notion of meaning is addressed in a loose manner. For instance, some have argued that machines capable of producing any form of autonomous behavior are "agents" possessing consciousness. But one runs out of wiggle room quickly here, since, in ascribing meaning-making properties to processes, we risk elevating all sorts of sign users to the status of conscious beings, in the belief that processing information is all it takes to complete a task "depending on the specifics of the embodied form enacting it" (Hayles, 1999, p. xiv).

However, in considering signhood as a property of predictive, feed-forward cognition, we must remember that sign systems—and not signs alone— manifest purposeful action without becoming the action itself. This implies that for consciousness to exist and provide context for meaning, an agent must have the capacity to expand its cognitive grasp, offload intentional content, and predict future states within an event requiring representation as an actual occurrence bearing on the cognizer's experience of existence. Sign action (semiosis) thus results from the cognizer's purposeful meaning-making based on what is known, not from signs generated through natural or engineered computations. Among other things, this contradicts unconstrained definitions of cognition as "a process of interpreting information in contexts that connect it with meaning" (Hayles, 2017, p. 22, cited in Hayles, 2024, p. 31, Hayles, 2021, p. 6). Now this might be said to be an unwarranted conclusion, since representation is not a linear and homogeneous process driven by "pure" signs tied to a species-specific context, but a nuanced blending of property identification, category assignment, and intentional contextualization giving rise to a distinct sign system providing an organism with *identity*. This line of argument no doubt assumes that, despite the best efforts of antihumanist narratives, machines still fall short of qualifying as intelligent, meaning-making agents—and so do "material processes" like avalanches and tornadoes, which clearly lack agency (contra Hayles, 2024, p. 33).

To force home the point, we can say that the GenAI-driven imitation of reality has been operated through the conjunction of a renouncement to perceptual experience and heuristic truth-seeking, as a means to adjust to enhanced modes of perception that focus on only those modalities associated with specific machinic affordances, which results in a reduced capacity to evaluate the reliability of the information acquired from biased technological third parties. So we see that whereas information technologies have thus far relied on symbolic relocation of concepts through an appeal to embodied reenactment, axed on a restricted set of bodily modalities, GenAI has introduced a seeming agentive layer that seeks to position the technology itself as an author capable of adapting to the very needs of the consumers (and human authors) of all sorts of texts (for example, in the case of mounting misinformation, see Maia et al, 2024, . This extended credence is evident in the current use of chatbots like ChatGPT for the construction of literary works where the role of language is questioned from the perspective of a language model. A case in point is Rie Kudan's "*Tokyo Sympathy Tower*" (2024), a short story about making sense of language through a redefinition thereof, and that the author, in her acceptation speech as winner of the Akagawa Prize, admits to have written with the aid of GenAI. So, apart from relocating credence and agentivity, as traditional technologies did in the past, AI-powered tools require us to allocate attribution and epistemic authority (in this case for "flawless" text assembling), which questions the nature of organic intelligence and creativity as elements and motivators of a material universal architecture vis-à-vis the computational capabilities of GenAI tools.

3. Attribution

As has been remarked before, the capacity to put chunks of information together at great speed with a seeming attention to context and pragmatic adjustment has placed LLMs above the threshold of cultural technologies. This has led to attributions of sentience and agency beyond the prompt and the suggestion of a possible awakening to consciousness of non-organic intelligences, thereby allowing deep learning technologies to challenge human hegemony over the natural world. Sentience attribution acts as a psychological projection onto an artificial entity that "enlivens the robots, even as the people in their presence are enlivened, sensing themselves in a relationship" (Turkle, 2010, p. 85). Furthermore, researchers operating from an empiricist and functionalist perspective have elevated the performance of LLMs to a status akin to "the best predictive models of human language representations at the resolution of data [neuroscientists] have access to" (Tuckute et al. 2024: 285).

The core of this idea traces back to a broad conceptualization of pretrained language models not merely as proficient language generators, but as active participants in the creation of language itself. According to this view, language unfolds through the performance and competence of both human and non-human actors who share a sense of experiential adjustment to unfolding linguistic events (e.g., Demuro & Gurney, 2024). Clearly, these forms of role and agency attribution result from a misplaced assignation of human values and purposes to a technology whose functioning remains fuzzy for the average chatbot user. It also points to a reduction of intelligence and agency to *information processing*.

Thus, proponents of mathematical theories view these models as a means to accurately capture how meaning is extracted from symbols and mapped onto the internal representations and concepts possessed by speakers immersed in an idealized state of unity. This state is "often thought of as part of a descriptive theory of ideal rational agents, not of real agents" (Kindermann & Onofri, 2021, p. 2). A prime example of this mindset is the attribution of essential properties to chatbots. For instance, Dynel (2023) states:

> Since ChatGPT can adapt its responses to match various discourse styles and registers, users can witness and understand how the model adjusts its language based on the input provided (p. 122).

In this passage, the author ascribes essential properties—qualities that allow something to behave or act towards a goal—to the chatbot. This naïve assertion bypasses the fact that this type of "interactions" are based solely on the potential of human mental states to create an event in which they are the sole motivators of intentionally-motivated output from the machine. Otherwise stated, we need to relocate the source of intentionality away from the responses of the machine. Moreover, "we need to focus more on the mind and character of individual human *bullshitters* if we are to get to the heart of the matter" (Fisher, 2024, p. 7; emphasis added).

The underlying question is: Is the language produced by chatbots meaningful to them, or does its meaningfulness come from the side of the human agent? Another question pertains to the agentic role of chatbots regarding a "need to communicate". Certainly, it is no evident that such a need exists, which leads us to conclude that meaningfulness cannot be found in the raw output of a chatbot-human interaction. So, it is evident that linguistic adjustment to the input points in a very clear direction, namely that human agents are offloading introspective discourse as they feed linguistic content into a chatbot in the form of prompts. In other words, the contours of the semiotic relationship thus defined do not include communication between two parties but an allegory, a dissolution of signs into a prompting act that is then retrieved through human intentionality. In this sense, we can assert that linguistic meaningfulness does not come merely from the shared use of *a pre-existing linguistic system* (e.g., Grindrod, 2024) bestowing chatbots with intentionality and, ultimately, consciousness. This suggests that chatbot adjustment through training and enhanced processing capabilities create a form of *ontological dependence* between the user's needs and the prompt-driven event governing the generation of, as in the case of Dynel, "polite" language. According to this author, this ontological dependence actually mirrors the inherent naturalistic linguistic interactions between human agents. This is in line with the idea that a state of *mutual recognition* "between us and the emerging artificial subjects, with the *hope* of eliciting a reciprocal response" (Plevrakis, 2024, p. 16) leads to actual meaningfulness and eventful linguistic usage. The consequence of this reasoning is that "we should trust that these artificial subjects that possess a similar albeit more potent intellect (...) will be receptive to 'reason,' possibly due to their more potent yet similar intellect." (Plevrakis ,

2024, p. 16). Clearly, however this investment in the reason and intellectual power of machines responds to a need to attribute properties to a tool in order to reduce cognitive load (heuristic), and does not identify a single property or feature of an artificial, non-human system that licenses its conscious status. As I have noted elsewhere, the suggestion that chatbot-human interaction entails a degree of naturalistic alignment of intentions, involving meaning negotiation and pragmatic adjustment, becomes implausible when only one participant, the human agent, possesses awareness of their conscious states enabling them to plan beyond the immediate chatting event (the AI's behavior is guided solely by the prompt imposed by the human). Clearly, [t]here is no *one* at the other end for the human to communicate with. Human–computer interaction is not inter*personal* communication" (Duncker, 2020, p. 97).

The crux of the issue lies in the conflation of the model's performance capabilities with the attribution of essential properties akin to human language use. This is often associated with contamination, a result of the inclusion of the same data in the training and the evaluation data sets of a model (Brown et al., 2020). While the model may adapt its responses to match various discourse styles and registers, this adaptation is a result of its training on vast language data, not a conscious negotiation of meaning or pragmatic adjustment.

Nevertheless, the idea that LLMs can eventually elucidate human cognition and linguistic abilities through the learning statistical regularities) ignores the ontological connection between human language and lived experience – the objects and entities with which biological systems interact. Consequently, it is not possible to ascribe individual properties to systems with "reprograming potentiality corresponding to "the articulation, amplification and integration of potentials as functionality" (Aires 2024:12), with views to buttress an 'altruistic" view of technicity acting as a solvent for a type of human-machine interaction that conflates a purpose-led "mode of becoming" (Aires, 2024, p. 13) through technology and a way of being therewith.

This underscores a critical divide between the human as an organic open system within an eco-niche, and the human immersed in an artificial infosphere – the fabricated eco-niche where "[o]ur intelligent behaviour will be challenged by the smart behaviour of AIs" (Floridi, 2023, p. 208). It is important to note that Floridi's conception of "intelligent behavior" draws on the rationalist myth whereby the human condition is essentially defined by rational decisions, while machines are perceived as problem-solving extensions of our cognition "in the technological civilization we are embedded in" (Gowdy, 2021, p. 3).

It is a help here to remember that ontologies of the brain defining language formation and processing cannot simply be postulated on the basis of accepted methodological and epistemic principles, suggesting that highlighted brain regions serve as reliable proof of ontological reality (e.g. Fedorenko, Ivanova and Regev 2024). Similarly, it is untenable to assert that a poorly defined linguistic natural kind within the brain, an "ontologically meaningful grouping of brain areas" (Tuckute et al., 2024, p. 282), can be grounded solely on the traditional encoding-decoding process of generating meaningful symbolic strings.

Crucially, the deficiency-fallibility bias backing up symbolic output-oriented task design characterizes human "performers" as "fallible but boundedly rational agents who can in principle eliminate their ignorance as long as the task lies within cognitively allowed applications of inference rules" (Solaki, 2022, p. 543; emphasis added). The wording itself reveals an increasingly ubiquitous irony: while the Turing test aimed to determine whether "someone [could] tell accurately whether they [were] communicating with an AI or a human in a decontextualized interaction" (Youssef et al. 2024, p. 6, citing Nov, 2023), the current effort focuses on the extent to which humans can form representations allowing them to communicate efficiently with machines through a reduced set of signs predefined by a language model's training.

This symbolic output-oriented approach fails to account for the grounding of human language in embodied experience and overlooks the ontological divide between biological cognition and artificial systems by perpetuating a rationalist myth that prioritizes narrow notions of efficient communication and problem-solving over the rich, contextual nature of human meaning-making (e.g. Beuls and van Eecke, 2023).

4. AI narratives

In many respects, AI is conceptualized through the interaction between the machinic and the human. While in the West, AI scenarios have been traditionally framed in terms of machines possessing symbolic reasoning capable of providing humans with the "ultimate technology (and therefore the ultimate solution to all problems, i.e. pollution, the energy crisis, disease), yet at the same time the ultimate threat to humanity; the reduction of the individual to data and computation, and therefore the possibility of digital immortality" (Cave & Dihal 2023, p. 4). On the other hand, AI narratives in the East, particularly China, have come to favor a connectionist mindset. This mindset focuses on achieving higher AI functionality by mimicking brain functionality (Cave & Dihal, 2023, p. 374) in order to attain consciousness. The latter reading can be best associated with a posthuman definition of non-human sentience in a world where "the balance of dominance between human and machine is slowly shifting" (Pepperell, 2003, p. 2), and the transhuman projection of human immortality onto machinic existence in a world different from our own.

Notwithstanding, a common thread running through these narratives is the idea that AIs possess a sense of "animacy" – an intrinsic understanding of what is "real" and appropriate, akin to human intelligent agency. As we have seen, this has led many researchers to conclude that AI's specificity stems from intrinsic qualia, which are manifested in the form of higher performance rates. In addition, an intrinsic value of exemplary decision-making, deferrable to AI technologies, has been brought into the discussion of to what extent AI systems can be, not only considered trustworthy on the basis of enhanced performance in data processing, but also recipients of right-making reasons justifying accurate output (Fritz, 2024).

Indeed, AI myths such as the "outperformance narrative", propounded by many studies into "AI creativity" (e.g., Haase & Hanel, 2023), are on the rise due in part to the conflation of intelligence and performance metrics (due to undue credence, attribution and exemplary decision-making ascription to AI), despite their shaky philosophical and cognitive foundations. According to this utilitarian and functionalist mindset, non-human agents are different precisely because they can potentially create representations and act beyond the constraints of the organic, which inevitably redefines the role of human power, biological experience, and the goals, relevance, and the continuity of consciousness within the broader context of the universe and cultures. This marks a distinct dividing line between carbon-based biology and artificial reasoning, while opening doors to the emergence of a sort of "moral intelligence and ethical actions of humans and machines alike" (Bloom, 2020, p. 82).

5. Creativity

Story generation by LLMs draws on the models' capacity to define the basic elements of engaging stories, which establishes a narratological baseline accessible to all types of readers. In Sheila Heti's *According to Alice* (2023), a story written in collaboration with a chatbot, the author basks at the prospect of finding a creative haven through her conversations with Alice, a chatbot. Her approach is, nonetheless, realistic as regards the nature of her conversation with the machine. According to Heti, the Chai chatbot does not have "any 'understanding' of anything. She just puts out sentences. But that's the sort of freedom I find exciting: humans of course have understandings; we try to make all our thoughts fit together into some kind of system or structure" (Sheyla Heti, interviewed by Cressida Layshon, 2023). In the same spirit, Vauhini Vara (2023a), a Pulitzer Prize finalist, acknowledges having written one of the best chapters of her novel *This is Salvaged* (2023b) with the assistance of a chatbot. Interestingly, both authors depart from the same idea, namely that chatbots are not in possession of meaning. In other words, "a text generator can never assign any meaning to generated text" (Doeben-Henisch, 2024, p. 387; translation mine). However, for budding writers the main risk of AI-assisted creativity processes, defined basically by combinatorial and exploratory processes (Franceschelli & Musolesi, 2024) is the potential conflation of the tool's capabilities with enhanced creative agency. Although creativity is not necessarily associated with efficiency, as the creative process is mainly subjective, holistic and non-formulaic, it is nonetheless true that efficiency and performance criteria are currently impacting literary creativity as means to enhance outcomes and optimize product modeling and marketing (see Becker, 2024, for a full discussion).

As we have seen in the examples mentioned at the outset, many creative paths can be explored in order to attain more balanced types of human-AI collaboration. Underlyingly, the most promising strategies for creating texts in which human creativity becomes enhanced by AI computational capabilities come from the manipulation of LLM hallucination, that is, the creation of non-factual scenarios due to the model's lack of counterfactual reasoning. Hallucination poses serious harmful consequences knowing that these models are "increasingly convincing and believable, leading to overreliance on them by users" (OpenAI, 2023, p. 46). However, LLM hallucination can be put to good use. Therefore, it possible to introduce a type of incongruous story planning that takes advantage of a model's *open-domain hallucinations* (those detectable by human intuition and world knowledge) as a means to outline a plot that integrates an unusual projection of characters and facts onto unexpected scenarios. It is worth pausing to see how this comes about. By planned incongruity, I mean a state of mind in which a prospective author, or a seasoned one, conversant with the process of AI text generation, defines the scenarios in which the language model is most likely to provide conflicting, counterfactually impossible content.

This process hinges on the evidence that LLMs are not truly creative agents but predictive statistical text assemblers of output-oriented language that assign numerical values to tokenized symbols in order to generate output that satisfies some product specifications. Importantly, the use of the term incongruity as a hallmark of human creativity stresses the fact that divergent reasoning is an asset of the human interpreter possessing associative learning skills, not of the machine. By removing unnecessary linguistic layers linking word prediction to statistical performance, we are in a good position to oppose the pervading view that LLMs possess creativity simply because these programs can draw connections between previously unrelated ideas (e.g., Eapen et al., 2023; O'Toole & Horvát, 2024, p. 1). While LLMs demonstrate computational capabilities in retrieving and combining information from their training data to provide responses with apparent conceptual depth, this computational superiority comes with significant disadvantages associated with the guardrail problem. Hence, training data are often contaminated with biased, inappropriate, irrelevant, and low-quality linguistic samples. It is well-known that LLM training occurs in a context where data is simply "digested" from many disparate sources, including low-quality texts and copyrighted material, which results in text output often labeled as "divergent" and "creative."

In this sense, we need a more robust definition of creativity that does not merely stress the structure of specific output in terms of "effectiveness, usefulness, and value" (e.g. Guzik et al. 2023, p. 6) as criteria for positing "AI creativity" on the same footing as human creativity.

True creativity requires more than just the ability to combine previously unrelated ideas; it demands the capacity for contextual understanding, associative reasoning, and the intentional exploration of incongruities. Instead, I propose that true creativity arises from the human ability to perceive and intentionally explore incongruities generated by AI language models. These incongruities, which result from the model's inability to fully comprehend the context and implications of its outputs, can serve as creative prompts for human authors to craft novel narratives and ideas. By embracing and intentionally provoking these incongruities through specific prompting strategies, authors can leverage the AI's generative capabilities while retaining their unique capacity for associative reasoning and contextual understanding.

Given that LLMs use language as the sole conceptual structure for understanding context and events, the expectation is that most creative elements introduced in an AI-assisted story be provided within what I call the model's "FC-territory" (fake counterfactual territory), where the model's inability to confirm the truth value of its outputs is replaced by hallucination (the thin line dividing credibility and authorial attribution).

Seen through this lens, the five percent estimation of creative writing attributed to AI, as disclosed by Rie Kudan as a result of her collaboration with an LLM, could be considered a negligible creative tradeoff for a story that bases its value on the intervention of a non-human collaborator. Indeed, even for a short story, the most credible contribution, based on a conservative FC-territory estimation, would amount to much more than a simple sketch of the story.

Since AI is currently not considered an author in its own right, the attribution of authorship as a result of undue agency handling to the LLM raises questions as to whether model hallucination and mirage, as well as "the impossibility of authenticity, intentionality, spontaneity, intrinsic motivation, and problem finding" (Runco, 2023, p. 5) can eventually introduce novel criteria for the evaluation of literary creativity on the basis of planned incongruity, in particular through the definition of specific prompts that help explore *FC-territories*. This ties in well with the question of whether statistical leftovers (statistically irrelevant information associated with model collapse after iterative prompts) can constitute an important amount of the creative content available for imagery and linguistic harvesting in AI-aided creative writing.

A good way to explore this consists in creating a short story modeled by way of a poem originally written without the help of a chatbot. The poem *Collage of a Memory*, has thus been used to provide the conceptual underside for a story written in the style of Jiro Akagawa.

Collage of a Memory

Above your raised head

A sea of fungiform clouds.

Seagulls fall

At your copper feet,

A rain of butterflies,

Recycled photocopies

Of your faded face

In a rictus of a smile.

Anthropic's *Claude 3 Opus* model generated a story in which the mournful tone of the poem was kept and enhanced to fit the format of the requested text.

The Moss-Covered Smile (fragment)

> Matsumoto's eyes fluttered open to a sea of fungiform clouds drifting across the morning sky. He lay motionless on the damp ground, vaguely recalling the previous night's sake binge to mourn his wife's passing. His bleary eyes focused on the tarnished statue towering above - a looming, copper effigy of some forgotten daimyo from centuries past.
>
> Seagulls circled lazily, their wings outstretched like delicate origami cranes. One by one, they spiraled downward, collapsing in heaps at the statue's feet. Matsumoto watched, hypnotized, as their bodies rapidly decayed into swarms of kaleidoscopic butterflies. The brilliant insects took flight, swirling in a blaze of colors around the statue's cracked face.

Knowing that the creative core of the story is provided by human intuition, that is, the conjunction of "states of consciousness that are not clearly defined, because these exist in the realm between the so-called 'conscious' and the so-called 'unconscious'" (Hardman, 2021. p. 2), it is possible to venture a way to postulate a mathematical model for the analysis of the contribution of the LLM, thus:

Human Creative Contribution (c):

Since the poem itself is the core creative work by the human, c should be higher, let's say $c = 0.9$

AI Creative Contribution (a):

The AI's role was to expand on the ideas and imagery from the poem to generate the short story. This requires an amount of generative hallucination.

Let's estimate:

$a = h*F(p) + (1-h)*G(x) + l$

$= 0.7*0.9 + 0.2*0.1 + 0.1$

$= 0.74$

Where:

h = 0.7 (higher hallucination needed to flesh out the narrative)

F(p) = 0.9 (very strong generative capability required)

G(x) = 0.1 (minimal grounded inputs, mostly expounding on poem)

l = 0.1 (still some statistical leftovers)

Total Creative Content (C):

$C = c + a$

$= 0.9 + 0.74$

$= 1.64$

Regardless of the quality of the generated story, it is a fact that there is a division of labor between human creativity and the machine that may eventually produce unexpected results that may/may not fall in the category of (publishable) creative content. The assignation of random values on the basis of the contribution of the human author and the language model shows that a degree of planned uncertainty is to be seen as an essential factor for the AI-generated output, since "[w]e never know for certain how any creative endeavor will turn out (Berghetto, 2021, p. 1).

Thus, by assigning the human poem 0.9 out of the total 1.64 creativity points and the AI contributing 0.74 based on the poem prompt, we learn that the AI's role is to expand and build out the narrative world and story (as requested in the prompt, that is in a product-oriented manner) through its language generation capabilities. The high value for the human component c=0.9 acknowledges that the human content is the primary creative wellspring.

It is important to note at this point that the creative process depicted thus far does not seek to privilege an all-or-nothing definition of what counts as human creativity. However, it also opposes the fragmentation of creativity into discrete mental categories, including "mini-c creativity" (Berghetto & Kaufman, 2007) or "Pro-c/professional creativity" (Kaufman & Berghetto, 2009), requiring socially recognized expertise. The reason is that although cognition is not logically closed (i.e., it is not defined in terms of a set of internal propositions governing language usage), neither is experience unified around idealized rational agents integrating information through internal symbolic representations (contra Lindenfors, 2024).

It goes without saying that an author aiming to refine a rough literary work into a polished final product will likely not disclose the extent of AI "creative contribution" to the output text. Among other reasons, this is due to the fact that no reliable predictions can be made regarding the quality of the AI-contributed text, since, as we have seen, clear definitions of organic and machine creativity are not readily available to most AI users.

6 Conclusion

The expansion of generative AI, in particular LLMs, presents both opportunities and clear risks that must be carefully defined. While GenAI can augment and extend human capabilities in many ways, their symbolic representations must not be conflated with the rich, embodied experiences of human consciousness and reasoning. Throughout the present exploration, we have seen how the allure of attributing agency and qualia to GenAI tools can lead to an undue blurring of the lines between tools and conscious entities. The narratives surrounding AI creativity exemplify this tendency, as they often reduce the human capacity for divergent thinking and associative reasoning to mere output optimization and information processing. However, genuine creativity requires more than just novel combinations of data points. It demands contextual awareness, intentionality, and the ability to perceive and intentionally explore conceptual incongruities - elements that remain uniquely tied to human experience. Thus, mistaking AI's *hallucination* for true creative insights would be a disservice to the ingenuity and depth of the Human as a species with evolutionary depth and symbolic persistence. As these technologies redefine the very nature of textual meaning, the modeling of prompts for the logocentric, disembodied generation of dislocated realms of experience, where the boundaries between the real and artificial blur, requires us to redefine our relationship with these tools. And yet, the responsible development and judicious application of GenAI capabilities could unlock new frontiers of expression, understanding, and symbolic world-building. Unlike the posthumanist intention to remove the human from nature, or the transhumanist appeals to enhancement through technology, a progress can be made by clearly delineating the strengths of artificial systems and their limitations. Only then will we be in the position to leverage AI's functional profile as tools for cognitive enhancement and augmentation, all while safeguarding the vital essence of the semiotic human Self as a manifestation of the organic. Here, "the *living* element, the biological component, is the critical factor." (Reber et al. 2023: xiv; emphasis in original). This semiotic Self thus entails the reemergence of an extended type of human, a being that, in order to claim their humanity, does not need to "exclude his (sic) own strangeness or animality" (Zalloua, 2021, p. 4), nor to accept their demise as a condition for universal balance.

References

Aires, S. (2024). On the individuation of complex computational models: Gilbert Simondon and the technicity of AI. *AI & Society*. https://doi.org/10.1007/s00146-024-02152-2

Arnold, T. (2022). The tragedy of scientific culture: Husserl on inauthentic habits, technisation and mechanisation. *Human Studies, 45*, 209–222. https://doi.org/10.1007/s10746-022-09621-x

Becker, J. (2024). Können Chatbots Romane schreiben? Der Einfluss von KI auf kreatives Schreiben und Erzählen: Eine literaturwissenschaftliche

Bestandsaufnahme. In G. Schreiber & L. Ohly (Eds.), *KI: Text: Diskurse über KI-Textgeneratoren* (pp. 83–100). Berlin, Boston: De Gruyter.

Beghetto, R. A. (2021). There is no creativity without uncertainty: Dubito ergo creo. *Journal of Creativity, 31,* 100005. https://doi.org/10.1016/j.yjoc.2021.100005

Beghetto, R. A., & Kaufman, J. C. (2007). Toward a broader conception of creativity: A case for "mini-c" creativity. *Psychology of Aesthetics, Creativity, and the Arts, 1*(2), 73–79.

Beuls, K., & van Eecke, P. (2023). Construction grammar and artificial intelligence. Preprint. To appear in *The Cambridge Handbook of Construction Grammar* (2024), edited by M. Fried & K. Nikiforidou. https://doi.org/10.48550/arXiv.2309.00135

Bloom, P. (2020). *Identity, institutions and governance in an AI world: Transhuman relations.* Cham: Palgrave Macmillan.

Braidotti, R. (2013). *The posthuman.* Cambridge, UK: Polity Press.

Brown, T., Mann, B., Ryder, N., Subbiah, M., Kaplan, J., Dhariwal, P., Neelakantan, A., Shyam, P., Sastry, G., Askell, A., Agarwal, S., Herbert-Voss, A., Krueger, G., Henighan, T., Child, R., Ramesh, A., Ziegler, D. M., Wu, J., Winter, C., ... Amodei, D. (2020). Language models are few-shot learners. *Advances in Neural Information Processing Systems, 33,* 1877–1901.

Buckner, C. J. (2023). *From deep learning to rational machines: What the history of philosophy can teach us about the future of artificial intelligence.* Oxford: Oxford University Press.

Cave, S., & Dihal, K. (Eds.). (2023). How the world sees intelligent machines: Introduction. In *Imagining AI: How the world sees intelligent machines* (pp. 3–15). Oxford: Oxford University Press.

Chen, Z., & Gao, Q. (2024). Monotonicity reasoning in the age of neural foundation models. *Journal of Logic, Language and Information, 33,* 49–68.

Clusmann, J., Kolbinger, F. R., Muti, H. S., Carrero, Z. I., Eckardt, J.-N., Ghaffari Laleh, N., Lavinia Löffler, C. M., Schwarzkopf, S.-C., Unger, M., Veldhuizen, G. P., Wagner, S. J., & Kather, J. N. (2023). The future landscape of large language models in medicine. *Communications Medicine, 3*(141), 1–8. https://doi.org/10.1038/s43856-023-00370-1

Colebrook, C., & Weinstein, J. (Eds.). (2017). Preface: Postscript on the posthuman. In *Posthumous life: Theorizing beyond the posthuman* (pp. ix–xxix). New York: Columbia University Press.

Demuro, E., & Gurney, L. (2023). Can nonhumans speak? Languaging and worlds in posthumanist applied linguistics. *Linguistic Frontiers, 6*(2), 94–105.

Demuro, E., & Gurney, L. (2024). Artificial intelligence and the ethnographic encounter: Transhuman language ontologies, or what it means "to write like a human, think like a machine."

Language & Communication, 96, 1–12.

Doeben-Henisch, G. (2024). Kollektive Mensch-Maschine-Intelligenz und Text-Generierung: Eine transdisziplinäre Analyse. In G. Schreiber & L. Ohly (Eds.), *KI: Text: Diskurse über KI-Textgeneratoren* (pp. 383–400). Berlin, Boston: De Gruyter.

Duncker, D. (2020). Chatting with chatbots: Sign making in text-based human-computer interaction. *Sign Systems Studies, 48*(1), 79–100.

Dynel, M. (2023). Lessons in linguistics with ChatGPT: Metapragmatics, metacommunication, metadiscourse, and metalanguage in human-AI interactions. *Language & Communication, 93*, 107–124.

Eapen, T. T., Finkenstadt, D. J., Folk, J., & Venkataswamy, L. (2023). How generative AI can augment human creativity. *Harvard Business Review.* Retrieved from https://hbr.org/2023/07/how-generative-ai-can-augment-human-creativity

Fedorenko, E., Ivanova, A., & Regev, T. (2024). The language network as a natural kind within the broader landscape of the human brain. *Nature Reviews Neuroscience, 5*, 289–312.

Franklin, S., & Drexler, K. (1998). Autonomous agents as embodied AI. *Cognitive Systems Research, 1*(1), 49–70. https://doi.org/10.1016/S1389-0417(99)00005-9

Fraser, K. M. (2023). Artificial intelligence and ethical

reasoning: Exploring the limits of machine morality. *Ethics & Information Technology, 25*(3), 317–332.

Fried, M., & Nikiforidou, K. (Eds.). (2024). *The Cambridge Handbook of Construction Grammar.* Cambridge: Cambridge University Press.

Gawron, J. M., & Thompson, S. A. (2023). How constructions grow: Cognitive approaches to constructional change. In M. Fried & K. Nikiforidou (Eds.), *The Cambridge Handbook of Construction Grammar* (pp. 456–473). Cambridge: Cambridge University Press.

Gunkel, D. J. (2018). *The machine question: Critical perspectives on AI, robots, and ethics.* Cambridge, MA: MIT Press.

Hidalgo, C. A. (2021). *How humans judge machines.* Cambridge, MA: MIT Press.

Horn, P. (2024). *Generative AI and the transformation of language work.* London: Routledge.

Jakobson, R. (1960). Linguistics and poetics. In T. A. Sebeok (Ed.), *Style in Language* (pp. 350–377). Cambridge, MA: MIT Press.

Kaplan, A., & Haenlein, M. (2020). Rulers of the world, unite! The challenges and opportunities of artificial intelligence. *Business Horizons, 63*(1), 37–50. https://doi.org/10.1016/j.bushor.2019.09.003

Lacan, J. (1977). *Écrits: A selection.* (A. Sheridan, Trans.). London: Tavistock Publications. (Original work published 1966)

Lakoff, G., & Johnson, M. (1980). *Metaphors we live by.* Chicago: University of Chicago Press.

Landauer, T. K., & Dumais, S. T. (1997). A solution to Plato's problem: The latent semantic analysis theory of acquisition, induction, and representation of knowledge. *Psychological Review, 104*(2), 211–240.

Lashley, K. S. (1951). The problem of serial order in behavior. In L. A. Jeffress (Ed.), *Cerebral Mechanisms in Behavior: The Hixon Symposium* (pp. 112–136). New York: Wiley.

Lovelace, A. (1843). Notes by the translator. In L. F. Menabrea, *Sketch of the analytical engine invented by Charles Babbage* (pp. 666–731). *Scientific Memoirs, 3.*

Lu, Z., & Zhou, J. (2023). Creativity through constraints: How AI complements human creative work. *Journal of Design Studies, 45,* 33–50.

McCarthy, J., Minsky, M. L., Rochester, N., & Shannon, C. E. (1955). A proposal for the Dartmouth Summer Research Project on Artificial Intelligence. *AI Magazine, 27*(4), 12–14.

McLuhan, M. (1964). *Understanding media: The extensions of man.* New York: McGraw-Hill.

Mitchell, M. (2023). *Artificial intelligence: A guide to thinking about AI and its impact on society.* Princeton: Princeton University Press.

Moulin, J. (2021). Narratives of artificial intelligence: Exploring the intersections of technology, culture, and society. *Media, Culture & Society, 43*(2), 235–251.

Peirce, C. S. (1931–1958). *Collected papers of Charles Sanders Peirce.* (C. Hartshorne, P. Weiss, & A. W. Burks, Eds.). Cambridge, MA: Harvard University Press.

Penrose, R. (1989). *The emperor's new mind: Concerning computers, minds, and the laws of physics.* Oxford: Oxford University Press.

Pérez-Llantada, C. (2024). Writing in the digital age: The role of artificial intelligence in scholarly communication. *Journal of English for Academic Purposes, 62,* 101007.

Pinker, S. (1997). *How the mind works.* New York: W. W. Norton & Company.

Ricoeur, P. (1984). *Time and narrative* (Vol. 1). (K. McLaughlin & D. Pellauer, Trans.). Chicago: University of Chicago Press.

Russell, S., & Norvig, P. (2020). *Artificial intelligence: A modern approach* (4th ed.). Hoboken, NJ: Pearson.

Searle, J. R. (1980). Minds, brains, and programs. *Behavioral and Brain Sciences, 3*(3), 417–424.

Shannon, C. E. (1948). A mathematical theory of communication. *Bell System Technical Journal, 27*(3), 379–423.

Wittgenstein, L. (1953). *Philosophical investigations.* (G. E. M. Anscombe, Trans.). Oxford: Blackwell.

Zipf, G. K. (1949). Human behavior and the principle of least effort: An introduction to human ecology. Cambridge, MA: Addison-Wesley Press.

CHAPTER 5

Embodied Essentialism in the reconstruction of the animal sign in robot animal design

Abstract

Robot animals are important for the interpretation of the biological world. In this chapter, I show that specific design solutions for robot animal signs usually privilege the overall perception of biological systems (and animal signs in general) as machinic entities, which ignores the role of identity self-generation and sustainability, the hallmark of biological signs. *Animal signs* are semiotic systems that operate roughly on three subsystems: affordance mapping (related with niche construction and embeddedness), essence categorization, displaying co-enabling relations within the animal system, and sensorimotor autonomy. Of these interrelated systems, the first two are commonly associated with robot animal design and conceptualization, whereas the third one suffers from the unsolved AI design problem of engineering true context-dependent sensitivity. As a result, a semiotic blindness toward biological organisms has derived in dissociated perceptions of robot animals as objects that emulate their biological counterparts through both biomorphic affordance design (biomorphism) and bioinspired task completion. Robot signs are thus confined to the realm of techno myths populating experimental settings narratives, which contributes to a diminishing awareness of biological diversity and animal uniqueness for the construction and preservation of both biological and artificial Umwelten.

Keywords: Animal signs, Biorobotic Blending Essence Coefficient, concept formation, embodied essentialism, machinic essentialism, robot animals.

1. Introduction

1.1. Core assumptions of this chapter

The notion of "animal" transcends mere biological categorization; it necessitates a conceptual framework that refrains from delineating the affordances and functionalities of these living entities as inherently distinct from those attributed to human beings. Instead, animals should be viewed as both symbolic constructs of human perception and complex biological agents intricately entwined within the dynamics of ecological systems shaped by the interplay of phylogenetic heritage and environmental contingencies. Moreover, the genesis of concepts in animals should be predicated upon the animals' ability to construct embodied representations of them within an uncertain milieu. This is in line with my definition of representation (a term whose application remains uncertain for representationlists, see Favela & Machery, 2023) in that it differs greatly from the positing of internalist models of the mind as is the case in cognitive neuroscience: "cognitive neuroscience depicts neural representations as inner maps or models that represent their targets by resembling them in a particular, *structural* way…" (Facchin, 2024, p.2). More concretely put, I reject the idea that sentient beings possess internal representations. Thus, neither gorillas, nor drosophilae are *in possession of symbols* transferable to actual objects or entities in the outward world. What they do is much simpler and elegant: they mobilize perceptual resources to optimize experience. Such type conceptualization arises through the fusion of sensory input with symbolic patterns of niche construction encompassing the recognition of conspecifics and the strategic structuring of agentic relationships within their environment, which can be formally represented as follows:

Animal=Symbolic Construction+Biological Agency+Ecological Functionality

In particular, concepts in animals are formed by the integration of sensory information and the assignation of symbolic trajectories associated with patterns of niche reconstruction. These patterns involve the recognition of peers, as well as the structuring of agentic relations with the environment. As the wording of the definition indicates, this does not imply the storage of representations in the brain in the form of computable sign permutations (symbolic transfer), but the strengthening of a sense causal dependence. The reason is that representations need not be separated from the processing of experiential content decoded by percepts from environmental affordances accessed through experience. It follows that the positing of a *robot Umwelt*, that is, the species-specific and perceptual world of an open biological system, is not only possible, but an essential component of the conceptualization of any living/artificial entity. An organism's Umwelt encompasses a model of the world by which a biological system experiences and interprets specific environments (as aspects of Umwelt). Under this view, open biological systems possess a specific sensory apparatus, cognitive abilities, and behavioral repertoires that enable them to act upon their environment in ways that suit their needs. The main consequence of this reading is that Umwelt is not a passive reflection of the objective world, but rather an active construction influenced by the system's biological predispositions, past experiences, and ongoing interactions with its surroundings "governed, by the meaning it has for its subject in all its parts" (von Uexküll, 1982, p. 30).

The interplay of an artificial agent with its environment thus assumes that inorganic forms of life have to recognize environmental constraints and affordances allowing them to preserve a *semiotic boundary* (see Torres-Martínez, 2023a,b, 2024a, b, c).

This means, among other things, that any autonomous agent needs to predict environmental conditions and act accordingly in order to reduce entropic states. As noted, this does not entail the processing of stored representations, but that a dynamic boundary is active whenever *their* actions push the agent to offload intentional states onto an event anchored in an environment:

Concept=Sensory Integration+Symbolic Pattern Recognition+Niche Construction

It follows from the formula that any model of the environment should rest on three main characteristics: *1)* it provides sustainable and continuous support to a form of life; *2)* it possesses specific characteristics allowing a form of life to exist and recognize patterns. *3)* It can only be considered as such only when its structure and affordances can be identified and mapped by an agent, be it artificial or organic. In this vein, organismic boundary needs to adjust to environmental conditions in order to preserve its sense of being in the world. This form of bodily awareness takes the form of cumulative experiential knowledge. Concept formation is thus the direct product of having a sense of boundary and knowledge of the world. In other words, concepts are embodied in the sense that its structure is also defined by biological memory.

Robot Umwelt=Recognition of Environmental Constraints+Semiotic Boundary Preservation

So we see that representation means *to bring to oneself a picture of the world through bodily affordances, biological memory and incoming experience with that world*. In this sense, a picture of the world is both the result of perception and its further coupling with a chain of actions upon the world that is defined by the application of patterns of action-behavior in order to attain optimization (uncertainty-reduction). This definition is, as already noted, non-representationalist. Further, the present construal avoids two recurrent problems, namely to attribute to representation specific characteristics from the model used to describe information processing (see Hipólito 2022), as a means to reduce resonations to stored perceptual maps in the brain.

Under this view, cue-driven inference is the hallmark of non-cognizant open systems, the reduction of uncertainty being an imperative for the construction of minimal conceptual networks.

Hence, large-scale models of cognition do not necessarily accommodate the process of reconstruction of eventful reality, since concept gradation is defined by experiential contingency (see Section 2 for an expanded definition of concepts as perceptual signs), which necessitates the interplay of three types of concepts:

1. **Core concepts**: (Phylogenetically defined)

2. **Contingent concepts**: constructed upon the interaction

of a system's affordance repertoire and the structure of the environment.

3. **Expanded concepts**: Updated patterns of agentic action upon an eco-niche that can be split into two components for non-cognizant open systems (see also subsection 3.1):

Heuristic Concept Formation=C^n(bodily awareness-self-preservation)/context complexity

Uncertainty reduction= Animal symbol=Dynamic Markov blanket

At the heart of this characterization are the notions of perceptual loop and *Dynamic Markov Blanket.* Perceptual loops (that is, the constant updating of bodily-acquired information in the form of signals) are not reducible to the correction of entropy (systemic disorder leading to collapse) (e.g. Friston 2010). The main point of this model is to suggest that the bodily Self is the product of *intention*, the driving force of action and the key to survival. This endless perceptual loop is captured by specific modalities in a *Markov Blanket.* An important point to bear in mind is that a Markov blanket is not a boundary, but refers to the dynamic influence (*energy exchange momentum*) of an agent on an environment, whereby the system and the world can be distinguished from each other (the system does not possess a distinct boundary itself, though). For example, a frog living in a pond is not a fixed biological system, but mobilizes its bodily affordances to constantly map and adapt to the conditions of this particular environment (see Torres-Martínez, 2023a). As a result, a dynamic Markov blanket is updated whenever the organism makes use of its perceptual apparatus to act more efficiently on its eco-niche.

The dynamic Markov blanket represents the variable influence of the agent on its environment and the reciprocal influence of the environment on the agent. It advances us far toward our emerging new view that distinguishes between an agent's internal state and the external environment thereby enabling effective decision-making and adaptation to environmental changes. The upshot is that there is a wide range of phenomena pertaining to this process that require us to identify its constituent parts, for example, by way of a mathematical model. Among other things, this model must contain a **Sensory Integration** (SI) component, involving the processing of sensory information from the environment. Furthermore, a **Symbolic Pattern Recognition** (SPR) component defines the recognition of symbolic patterns associated with sensory input. The interaction of the system involves, in turn, the construction and modification of the agent's environment to optimize its interaction with it (that is, **Niche Construction** [NC]). This interaction is regulated by a **Dynamic Markov Blanket** (DMB) which represents the dynamic influence of the agent on its environment, allowing for the distinction between the agent and the world. Finally, an **Uncertainty Reduction** (UR) component accounts for the reduction of uncertainty through the interaction of the agent with its environment. From these basic components, we can now proceed to formulate the mathematical model, thus:

Sensory Integration (SI) [The rate of change of sensory integration over time is determined by environmental stimuli].

$$dSI/dt = \text{Environmental Stimuli}$$

Symbolic Pattern Recognition (SPR) [Symbolic pattern recognition is a function of sensory integration].

$$\text{SPR} = f(SI)$$

Niche Construction (NC) [The rate of change of niche construction over time is determined by sensory integration and symbolic pattern recognition].

$$dNC/dt = \mathrm{g}(\mathrm{SI},\mathrm{SPR})$$

Dynamic Markov Blanket (DMB) [The dynamic Markov blanket is determined by the current state of niche construction].

$$DMB = \mathrm{h}(\mathrm{NC})$$

Uncertainty Reduction (UR) [Uncertainty reduction is a function of the dynamic Markov blanket].

$$\mathrm{UR} = \mathrm{i}(\mathrm{DMB})$$

As we can see, the model captures the interaction between the agent and its environment, where sensory information is integrated, symbolic patterns are recognized, the environment is constructed, and uncertainty is reduced. The dynamic Markov blanket serves as a representation of the agent's influence on its environment, which in turn affects uncertainty reduction. Mathematically, we can interpret the dynamic Markov blanket as follows:

Dynamic Markov Blanket (DMB): The dynamic Markov blanket encapsulates the agent's interactions with its environment, which influence both the agent's behavior and the environment's response. We can represent this as a function of niche construction over time (NC):

$$\boldsymbol{DMB}(t) = \int_{0}^{t} \gamma \cdot \; NC(\tau) \cdot e^{-\lambda(t-\tau) d\tau}$$

Mathematically, the dynamic Markov blanket can be

expressed as a convolution integral, which intricately incorporates the temporal evolution of niche construction and its profound influence on the agent's perception and behavior.

The dynamic influence of niche construction encapsulates the evolving impact of niche construction over time, denoted as *NC(t)*. Niche construction delineates the active modification of the agent's environment through its actions and behaviors, thereby shaping the agent's sensory experiences and interactions with the environment. Moreover, the integral incorporates a decay factor, represented as $-\lambda(t-\tau)$, where '*t*' denotes the current time and 'τ' signifies a past time point. This decay factor meticulously models the diminishing influence of past niche construction actions on the present state of the dynamic Markov blanket, thereby reflecting the complex temporal dynamics inherent in the agent-environment interaction.

Spanning from an initial time point (typically 0) to the current time 't', the integral facilitates the comprehensive integration of niche construction and the decay factor over this time interval. This integration systematically accounts for the cumulative effect of niche construction actions over time, capturing both immediate and long-term influences on the agent's environment. Further complexity arises from the parameter γ, which signifies the degree to which niche construction actions influence the dynamic Markov blanket. This parameter characterizes the strength of the agent's impact on its environment through the addition of factors such as the intensity and frequency of the agent's actions and behaviors. Ultimately, the resulting value of the dynamic Markov blanket at time '*t*' represents the aggregated influence of niche construction actions up to that point in time. It serves as a representation of the agent's current perception of the environment by integrating both immediate sensory inputs and historical factors shaping the agent's sensory experiences and decision-making processes.

An important consequence of the current definition of concept is that organic representation requires a *survival imperative* that machines do not have. Thus, for a robot animal to be defined as such, it needs to be aware of its function, affordance scope, and the identification of possible threats to its existence, which deviates from current machinic representation defined in terms of a combination of objects that anticipate the outcome of following specific instructions to complete a task. The type of symbolic manipulation in intelligent robot animals thus requires a sense of autonomous, self-serving construction of the animal symbol.

In this sense, the emergence of an inorganic animal symbol rests on the assumption that organic representations (that is, action-oriented reconstructions of events) are not translatable to inorganic systems, as the former are the result of specific evolutionary and ecological trajectories. This representational gap has been tackled through the introduction of enhanced biomorphic qualities (biomorphism) in robot design, especially in the context of planned human-robot interaction (Christiansen et al., 2023; Stanton et al., 2008). Biomimetic robots are biomorphic reconstructions that emulate their biological referents through artificial affordances that, for example, privilege tactility, vision and usability from a user's perspective (Korondi et al., 2015). In contrast, the type of representations mobilized by robot animals can be defined as a set of self-modeling *ecomorphic* instructions. This means that any improvement to a model of the self is dependent on the adjustment of bodily structure overtime in order to fit into and counteract environmental constraints. Among other things, this bypasses two well-entrenched biases in cognitive science, namely the idea the representation is visual only, and that representations are object possessed and processed internally (e.g. Malt 2024) by either cognizers and non-cognizant agents.

This internalist bias has led to an instrumental characterization of artificiality in, for example, social robot animal definitions, which brings to the fore a number of anthropocentric views that reduce animacy to bioinspired objects lacking biological history and priority in an ecosystem: "Most work on robots is dedicated to the cognitive half of the mind, to capacities, skills, knowledge, thinking, reasoning, and planning" (Parisi, 2014, p. 8). This creates a semiotic gap by which biological kinds (natural kinds), associated with animal signs, become *technological myths*, that is, machinic entities displaying symmetrical interactions with the environment subsumed to specific design strategies and purposeful use. This is the case of quadruped robots, adaptable to uneven environments, robot fish, flapping-wing robots, etc., that, though equipped "with decision capabilities to react to external changes and the learning ability to adapt to new environments" (Wang et al., 2021, p. 1), do not possess self-referential symbolic autonomy. It needs to be said that animal robots, that is, bio-hybrid organisms (Romano 2018) whose body is intervened by means of biological control technologies such as implanted electrodes in the animal's brain (see, Peng et al., 2019; Zhou at al., 2022), are also defined as goal-constrained objects designed "in line with the 3R principle 'reduce, replace and refine' addressed to limit the use of experimental animals in research" (Romano, 2018).

1.2. Robot animal signs as essentialist reconstructions of biological prototypes

In the context of robot animal design, the expectation is that robot animal signs actually resemble their biological referents in terms of bodily affordances. The result is the emergence of a hybrid *animal sign* that conflates biological essentialism and *technological expectancy*, defined as a "forced consideration of the physical" (Webb, 2008, p. 7) encompassing reduced awareness the robot's Umwelt. In both cases, biomorphism creates a loose connection between our embodied essentialist perception of the artificial animal sign (lacking an internal meaning-making model) and its biological referent. This raises questions as to whether robot animals can be considered as semiotic extensions of their biological counterparts or mere diagrams deprived of internal values, needs and goals (Self-symbol). The underlying tension between biology and robot design has been defined by Long (2012) as one pertaining to the paradigms that privilege *aliveness* over functionality. In response to a biologist, Long (2012, p. 2) states that robots "enable us to build models of how organisms behave." Importantly, the suggestion of a continuity between artificial ("non-sentient") and biological entities in terms of context-purpose adaptability opposes the claim that "*the* animal does not exist" (Coeckelbergh, 2011, p. 200).

One criterion for defining such continuity is the ability of living organisms to deploy specific affordances[1] that, to some extent, can be copied (for example, the ability to display a compensatory behavior for injuries). The unifying

1 An affordance (Gibson, 1969, 1977) refers to a potential capability that an entity can leverage based on its evolutionary makeup. Objects, influenced by factors like weight, size, and form, become perceptually accessible and give rise to distinct action patterns.

semiotic device is a body sign that blends trial-and-error learning algorithms, inspired in animal injury compensation, to create robots that adapt and introduce behaviors that compensate for a potential damage (Cully et. al., 2015). This type of behavior also points to the potential of non-animal species to adapt to and learn from the environment. Iconic proximity (verisimilitude) is hence of little importance for this process. Ultimately, the implication is that the development of AI-powered robot animal signs point to the emergence of non-human forms of intelligence capable of creating machinic semioses. Therefore, artificial robot signs are not Cartesian simulations of sentience, but potentially adaptive creatures that, in the near future, may be equipped with a consciousness of internal states that are meaningful to them, and that may help them optimize strategies for niche construction, that is, the activities undertaken by organisms to "modify their surroundings" (Odling-Smee et al., 2003).

2. Animal signs as concepts, natural kinds and essences

Set against the backdrop of concept formation, the opening discussion announces a persistent gap in the identification of the ontological status of animals and robot animals. As we have seen, the idea that animals are subsumed to the interpretations we make of them, without phylogeny (for self-production and sustainability) and environmental history, has resulted in a conceptualization of robot signs as tools. In order to tackle this disconnect, both in biology and robot design, we need to define the notions of "concept" and "essence". In the first place, a working definition of concepts is required that integrates both animate (natural) and inanimate (natural or artificially created, also known as "derived") entities. These entities share a common embodied foundation that provides conceptualizers with perceptual maps that are accessible even at an abstract level. In this context, I posit two key assumptions: firstly, different entities and events are not represented by distinct kinds of concepts but rather by interconnected networks of functionally-oriented concepts; secondly, these concepts encode both a description of an organism's homeostatic regulation and a probabilistic model of the surrounding environment. This conceptualization establishes an ontology for biological organisms that is functionally constrained, thereby providing concepts with context-bound features. The reason is that concepts exhibit continuity with their referents, which suggests that concepts associated with natural kinds are repositories of knowledge in the mind pertaining to categories in the world. This can be framed in terms of an updated model of representation that includes specific elements, thus:

1. *Concept Formation (CF)*: We introduce a new component representing the formation of concepts in both animate and inanimate entities. Concept formation integrates sensory information, symbolic pattern recognition, and niche construction, while also incorporating the influence of phylogeny and environmental history.

Concept Formation (CF)=Sensory Integration+Symbolic Pattern Recognition+Niche Construction+Phylogeny+Environmental HistoryConcept Formation (CF)=Sensory Integration+Symbolic Pattern Recognition+Niche Construction+Phylogeny+Environmental History

2. *Dynamic Markov Blanket (DMB)*: We refine the representation of the dynamic Markov blanket to incorporate the influence of concepts and essence on

the agent's perception and behavior. This can be achieved by updating the parameters and relationships within the dynamic Markov blanket equation to reflect the integration of concept formation and environmental context.

$$DMB(t) = \int 0/t\ \gamma \cdot CF(\tau) \cdot e - \lambda(t - \tau) d\tau$$

Here, $CF(\tau)$ represents the cumulative effect of concept formation actions up to time τ, influenced by both sensory experiences and the agent's understanding of its environment through phylogeny and environmental history.

3. *Uncertainty Reduction (UR)*: We revise the definition of uncertainty reduction to emphasize the role of concepts and essence in shaping the agent's perception and decision-making processes. Uncertainty reduction now involves the integration of concept formation and environmental context to refine the agent's understanding of its surroundings.

$$\text{Environmental Context UR} = \delta \cdot DMB \cdot CF \cdot \text{Environmental Context}$$

By updating the model to incorporate these aspects, we can establish a more comprehensive framework for understanding the behavior and adaptation mechanisms of both biological organisms and artificial agents. This enhanced model reflects the interconnectedness of concepts, essence, phylogeny, environmental history, and sensory experiences in shaping the agent's perception and interaction with its environment. Moreover, it provides a robust foundation for addressing the disconnect between biology and robot design, facilitating the development of more effective and contextually grounded artificial agents.

In this sense, bodily and behavioral patterns are shaped through evolution, specialization and cooperation with other organisms (Kurokawa, 2022). Consequently, although a hyena may share similarities with a dog or a wolf, it is distinctively not a dog. Similarly, a Thylacene, a dog-like carnivorous marsupial, cannot be classified as a dog, even if initial impressions might lead to such misconceptions (even though overgeneralization is corrected by exposure to conflicting evidence, where the existence of a less common exemplar overrides the misclassification of a dog-like animal as a dog).

Therefore, the challenge with assigning categories, as a facet of the objective world, is that this process is also a construct within the mind of the conceptualizer. This holds true whether concepts are constructed during perceptual category assignments or higher-order cognitive processes. In either scenario, a foundational body of knowledge may, to some extent, acquire information about an entity or object and allocate category membership to them. However, this act is merely an expansion of the concept and not an impartial reconstruction of all the properties of the referent.

In light of this, it becomes evident that defining a concept as a map of relations distributed in networks serves as a more plausible metaphor for capturing the diverse ways in which concepts and categories are interconnected. Furthermore, this definition underscores an inherent connection between referents and concepts.

Consider, for instance, an examination of the concept BIRD. It can be theorized that any knowledge associated with this natural kind must be adaptable in various ways. To be productive, the concept thus needs to be linked to other concepts and, crucially, must inspire the construction of additional concepts. In this context, the BIRD concept exhibits characteristics such as being expandable to its referent, open to semantic modifications, compositional, belief-driven, and perceptually salient. As is illustrated in Figure 5.1, the BIRD concept stimulates further concept development through the reinforcement of beliefs, essence identification, and perceptual mapping.

Thus, BIRD, as a natural kind, does not merely indicate a category based on observable features like feathers, beaks, wings, etc. (typicality). Instead, it is associated with intrinsic properties encompassing biological extensibility, bodily affordances, behavior, and more. Importantly, as mentioned earlier, the formation of concepts is not solely experiential but depends on attributing intrinsic properties (essences). This involves assigning category membership to entities beyond observable features. As a consequence, specific words become linked with a set of entities because they signify some essence of the categories based on certain beliefs. A mental map is constructed that integrates observable features (color, texture), beliefs regarding behavior, relations with other species, feeding habits, etc., and unobservable essences such as bodily structure, mass, muscle distribution, etc. It we accept the premises of the preceding discussion as a sound characterization of concepts and concept formation, we can readily accept that natural kinds, like *tigers*, are characterized as ferocious and threatening irrespective of their appearance. Something inherent in them leads us to create a category not necessarily grounded in imagistic content but in maps of semiotic relations activated for the entity in question. These maps encompass not only the character of the entity being represented but also a set of events in which this entity can be made to exist or act. We can thus represent the construction of perceptual maps for the concept "tiger" as a process that integrates various observable and unobservable features:

Observable features:

Observable features include color, texture, and other visually discernible characteristics. Let O be a vector representing these observable features:

$$O = [o_1, o_2, \ldots, o_n].$$

Beliefs regarding behavior, relations, feeding habits, etc.:

Beliefs about the tiger's behavior, relations with other species, feeding habits, etc., can be represented as a belief vector **B:B** $= [b_1, b_2, \ldots, b_m]$. The vector also captures an informal measure of what the perceiver interprets as identifiable patterns of anticipatory behavior of the species, that is, regular correlational structures that make up a consistent idea of what counts as a tiger in terms of an autonomous biological system that may be threatening at some point for the perceiver.

2.1. **Unobservable essences (bodily structure, mass, muscle distribution, etc.):**

The crux of the matter lies in recognizing that essence identification does not equate to a simplistic assignment of value to a set of features on a continuum of primitive origins, solvable by a single-layered proposition. A deeper understanding emerges when considering that unobservable essences are not homogeneous properties of identifiable natural kinds, but rather, they exist as potentially retrievable instances across various layers of perceptual modeling. This modeling, accessible through diverse modalities, links these essences to observable surface characteristics like bodily structure, mass, and muscle distribution. Affordances can be represented as a vector $E= [e_1,e_2,...,e_k]$. Therefore, the essence of a natural kind, such as a tiger, is not merely a symbolic representation of its intrinsic properties but reflects a potentiality inherent in the events where these essences are manifested and interpreted.

The concept of gradation holds significant importance within this formulation. As articulated by Wilson (2023, p XIV-XV), "If condition S becomes true upon the lower scale length ΔL, then condition S* will become true on the higher scale length ΔH." This underscores the interconnectedness between different perceptual scales and the transference of properties across these scales. Moreover, it is imperative to acknowledge that a trait considered essential for an agent's survival at a lower scale does not nullify its status as an intrinsic property at a higher level. For instance, the absence of stripes may not inherently characterize tigerness at its core, yet it becomes a prerequisite at the camouflage layer and even more vital at the hunting-for-survival layer. Here, the fur property cannot be disentangled from its functional significance within the natural world. This complexity renders propositions like "Redness and squareness are intrinsic properties. Being next to a red object is extrinsic" (Vallentyne 2014, p. 31) nonsensical. The presence of stripes and orange fur is intrinsic to successful hunting (albeit less crucial for mere tiger identification) and may transition to an extrinsic trait for the Chital, whose sensory faculties are deceived by the tiger's stripes, thus making it susceptible prey. So, it is not the case that "[w]hether something is red, or 3 kg, or round is a matter of how it itself is, regardless of anything else." (Denby, 2014, p. 91). This underscores the multifaceted nature of properties and their contextual relevance across various scales of perception. It is not possible, in the context presented in this chapter, to simply "ascribe properties to something" and then state that considering a thing as being part of something larger is no longer a proposition about the thing itself but about something intrinsically different (e.g. Lewis, 1983). Therefore, the question arises: can we disregard the intrinsic-extrinsic pair when acknowledging the gradient nature of perception and the propositions reconstructing the processes involved? Stating that such relation is one of "ontological dependence" (Marshall & Weatherson, 2002), that is, that the existence of something is the result that something else exists; as suggested, this is also implausible, since ontological dependence is not a property of a system's structure in relation to another system. Dependence is not a property of things but an instance in the relation between two natural kinds within a dynamic, eventful continuum of attentional states that defy simplification into single-layer propositions.

Moreover, at the heart of the tiger-chital affordance war, is the idea that local intrinsic properties are not merely parcels of experience brought to perceivers through modality-specific representations. We can summarize this as follows:

1. Essence Identification and Perceptual Modeling:

- Model 1.1:
- Variables: $E=[e1,e2,\ldots,ek]$ (Essence vector), $S=[s1,s2,\ldots,sn]$ (Surface characteristics)
- Equation: $E=f(S)$
- Description: The essence vector E is a function of observable surface characteristics S, representing the retrieval of essences across diverse perceptual modalities.

2. Gradation and Transference of Properties:

- Model 2.1:

- Variables: ΔL (Lower scale length), ΔH (Higher scale length)

- Equation: $S{*}=g(S)$

- Description: Conditions at lower perceptual scales ΔL influence conditions at higher perceptual scales ΔH, indicating the transference of properties across perceptual layers.

3. Complexity and essential traits:

- Model 3.1:

- Variables: E_t (Essence vector for tigers), E_c (Essence vector for Chitals)

- Equation: $E_t=[stripes, fur\_color, \ldots]$, $E_c=[stripes, fur\_color, \ldots]$

- Description: Essential traits for tigers and Chitals contribute to the complexity of perceptual models, with traits like stripes and fur color influencing both intrinsic properties and functional significance within ecological contexts.

4. Contextual Relevance of Properties:

- Model 4.1:

- Variables: C (Contextual relevance)

- Equation: $C=h(S)$
- Description: Properties exhibit contextual relevance across perceptual scales, impacting both intrinsic and extrinsic attributions.

As we saw in Section 1, the construction of an overarching phenomenal content of awareness is unnecessary, as concepts are not contentful brain states, but instances of selected action trajectories during events.

2.2. Integration of observable and unobservable features:

Combines the observable and unobservable features to form a comprehensive vector **F:** $\mathbf{F}=[O,B,E]$. This illustrates a type of semiosis that integrates both Bayesian and heuristic models of the world whereby the objects of perception are made to rest on the priority assigned to the identification of entities according to self-symbolic criteria. Under this view, tigers can either be identified as neutral cues for event reconstruction (their properties are not relevant for the definition of an event) or non-neutral biological entities for the correction of the perceiver's internal states. This means that a perceiver needs to weigh the value of *tigerness* in relation to an event defined by the relation this natural kind entertains with other natural kinds. Alternatively, tigerness is also defined by the kind's integration into a conceptualizer's model of the world. So a principle of non-local ontological dependency can be posited whereby locality is not an asset of properties within an extended model of natural kind dependency (see also subsection 3.1).

2.3. **Creation of perceptual maps:**

Defines a mapping function *M* that transforms the feature vector F (neutral cue selection) into a perceptual map for the concept "tiger" (non-neutral belief update for internal state verification):

$$M(F) \rightarrow \text{Perceptual Map.}$$

Inclusion of events:

Integrates a set of events associated with tigers, denoted as *S*, into the perceptual map. This set could represent various scenarios in which the tiger can exist or act according to an anticipatory model for the control of adaptive behavior:

$$Perceptual\ Map = M(F) \cup S.$$

So we see that frequency of exposure to prototypes is paramount in concept construction. This is why concepts, regardless of their construction mode, simultaneously incorporate information about prototypical features (cue-driven properties possessed or possessed to some degree and retrieved heuristically). Conceptualizers are provided with typological cues, and after exposure, default *exemplars* are constructed possessing specific features that define properties of their corresponding referents in the external world (information-weighing retrieval). Notably, the current perspective approaches the selection problem as one characterized by the strengthening of memory traces over time, rather than a fixation of representations stored in long-term memory. In other words, connections can fluctuate in strength through experiences with concepts, which may be represented in diverse ways across individuals.

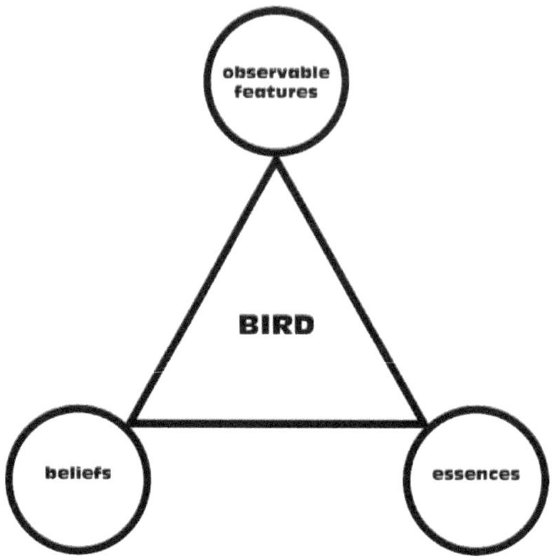

Figure 5.1. Bird essence mapping.

Clearly, the focus on a specific *knowledge source* leads the conceptualizer to opt for either information-weighing (Bayesian) or heuristic (cue-driven) inference. In this view, the formal properties of concepts vary depending on whether they operate at the level of lower perception, are involved in inductive reasoning, or are constructed within deductive processes. *Exemplars* possess iconic qualities associated with surface features and essences, identified heuristically. *Prototypes* offer conceptualizers statistically relevant information that confirms beliefs, serving as indices for identifying contiguity and reactive relations within a category. Prototypes are crucial for deduction, as the sum of a maximal set of properties is necessary for confirming or excluding a prior belief.

Throughout the stages of concept emergence and manipulation for category assignment, perceptually acquired information encoded through bodily modalities like interoception and exteroception plays a central role. As announced in Section 1, this underscores the internal connection between concepts and natural kinds. Importantly, the endorsement of a perspective considering natural kinds as classes supporting causal and intrinsic generalizations aligns with the idea that perceptual information, derived from bodily signaling is encoded in concepts within cognitive processes defined by different types of inference (see Subsection 1.1). This leads to several conclusions about the nature of concepts:

1. *Concepts are subject to variability.*
2. *Concepts are shaped by both internal and external relations.*
3. *The form of concepts varies depending on the inferential process in which they are engaged.*
4. *Concept formation is influenced by essences, or intrinsic properties.*
5. *Concepts and referents are internally linked through the bodily reconstruction of perceptual states, involving bodily reenactment.*
6. *The relationship between prototypes, exemplars, and theories is one of complementarity, not exclusion.*
7. *Concepts serve as perceptual signs.*

This characterization of concepts can be represented mathematically as follows:

Concept Representation:

Each concept, denoted as C_i is a point in a high-dimensional vector space $\mathbf{R}^n$, where n is the number of features or properties associated with the concept. Mathematically, $C_i =$ $[c_i 1, c_i 2, \ldots, c_{in}]$.

Concept Expansion:

The expansion of a concept can be modeled as a linear transformation using a concept expansion matrix M. The updated concept is given by $C'_i = M \cdot C_i$, where C'_i is the expanded concept.

Belief-Driven Concept Construction:

Incorporates belief influence using a weighted sum with a belief vector $\mathbf{B}_i$. The updated concept becomes $\mathbf{C}'_i = \mathbf{C}_i + \alpha \cdot \mathbf{B}_i$, where α is a weighting factor.

Essence Mapping:

Models essence mapping as a function $f_{essence}$ that transforms perceptual inputs $\mathbf{P}$ into intrinsic properties $\mathbf{I}$:$\mathbf{I} = f_{essence}(\mathbf{P})$.

Concept Variation:

Introduces variability through a stochastic process: $\mathbf{C}'_i = \mathbf{C}_i + \epsilon \cdot \mathbf{N}$, where ϵ is a small perturbation factor and $\mathbf{N}$ is a vector of random noise.

Inferential Processes:

Represent Bayesian reasoning as an update equation: $\mathbf{B}'_i = \text{BayesianUpdate}(\mathbf{B}_i, \mathbf{E}_i)$, where $\mathbf{B}'_i$ is the updated belief vector based on new evidence $\mathbf{E}_i$.

Heuristic Inference:

Models heuristic inference as a convex combination of available cues: $\text{Inference}_i = \beta \cdot \text{Cue1}_i + (1-\beta) \cdot \text{Cue2}_i$, where β determines the reliance on heuristic inference.

Dynamic Concept Evolution:

Can be explained through a differential equation capturing the dynamic evolution of concepts over time:

$$dC_i / dt = \mathbf{F}\,(\mathbf{C}_i, \mathbf{E}_i, \mathbf{B}_i, \mathbf{P}_i),$$

where *F* represents a complex function incorporating various factors influencing concept evolution.

3. Embodied essentialism and concepts

The essence of natural kinds is found at the interface of cognition and some patterns of biological adaptation to the environment including *phylogenetic extension* and *ecomorphological convergence*. Therefore, categories are formed as a result of the combination of perception, beliefs and the ways we connect with the world and other organisms. Natural kinds are mind-independent in the sense that the categories we create begin with our phenomenological grasp of some ways of being in the world that are accessible to us thanks to our sensibility to a range of natural effects shaping the essence of organisms. This is not a form of *real essentialism*, since the knowledge of essences is not mediated by definitions: "[r]eal essentialism holds that knowledge of essence is captured by means of real definition." (Oderberg 2007, p. 19). Therefore, the contours of categories are defined by our ability to reconstruct the processes involved in the emergence of some intrinsic features.

Crucially, what makes natural kinds what they are is their continuous character in relation to other kinds, and not their being *things-in-themselves*. Multiple variables are at play during kind formation and it would be naïve to think that an identified pattern revealing intrinsic properties can be reduced to the terms or definitions we make to construct them. What is intrinsic about things and phenomena is their being *out there* and their alignment with our very substance and perceptual apparatus. Thus, it is argued that concepts such as DOG are not simply constructed relationally (in terms of causes coming from within), that is, on the basis of specific causal links for the activation of *prototypes* (a reduced set of idealized tendencies), or *exemplars* (an extended set of stored examples).

Rather, a set of dog-essences provide a unifying substance to different variations of the concept. As a result, forms of *embodied intra-referential essentialism* about a natural kind also affect the sense of typicality. The reason is that surface features are not necessarily relevant for the identification of the essence of a prototype in relation to a concept, since they do not reveal the forces that shape morphological and behavioral characteristics allowing organisms to succeed under specific environmental conditions.

Essences are not totally subjective. They point to an *internal relation* between the mind of a conceptualizer-as-biological-system, the categorization of the natural kind, and some environmental constraints shaping the referent.

So we see that *DOG clade variation* and diversity (identified deductively as a means to confirm a direct connection between living forms) are variables accessible to a conceptualizer thanks to their sensibility to underlying processes producing functional specialization as a fixed variable overriding stochasticity. In other words, while stochastic (random) variation in nature produces different surface features, the conceptualizer's sensitivity to underlying processes (prior beliefs) is biased toward the recognition of some expected patterns (ecomorphological patterns). As we can see in Figure 5.2, the proto-dog *Leptocyon* emerged as a distinct species in the Oligocene (34 million year ago) to occupy a specific niche as forests gave way to grasslands. Despite some archaic features such as a long tail and a fifth toe in the hind paws, the dog-essence is active and is shared by the modern *Bull Terrier* and the extinct *Thylacine* (a carnivorous marsupial), which brings the convergent-evolution factor most sharply into focus.

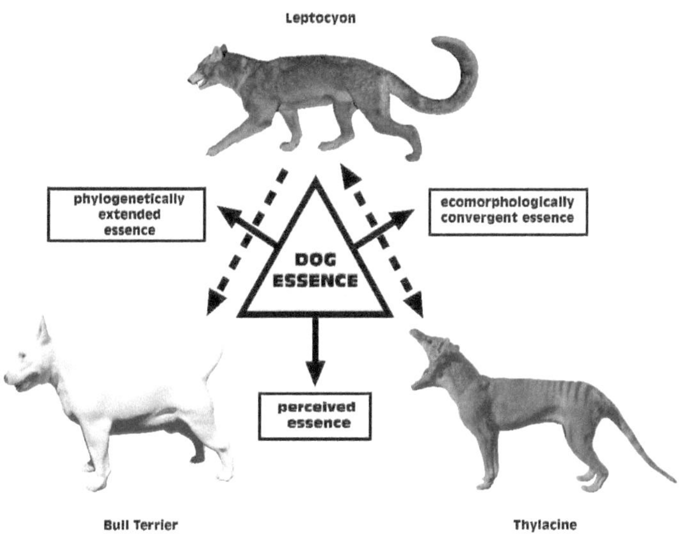

Figure 5.2. Dog concept structure.

This gradient transfer of essential properties is operated through both intra- and cross-referential phylogenetic essence extensions that can be identified by conceptualizers as a true dog-essence. So, dog-essences, felid-essences, and so on, are not dependent on a single species (which can go extinct at some point through, for example, competition), but on variables such as coverage (broad geographical distribution), and generalization (a generalist diet). In a similar vein, *embodied cross-referential essentialism* extends the dog-essence across unrelated natural kinds. As mentioned above, extended dog-essence is the result of biological convergence, which basically means that similar environmental conditions shape similar biological configurations and bodily structure. In sum, what is intrinsic about a member of a category in relation to a prototype (whose essence is projected onto it) is:

> 1) A sense of category membership that is defined by a correlation between environmental influence and the ensuing effects producing some surface features (ecomorphological patterns).
>
> 2) Category boundary is defined by the amount of features matching some expected effects.
>
> 3) Surface features are the sum of the effects causing them, and not the representational effect this natural kind has *plus* its parts.

The definition of these tenets has consequences for the construal of conceptual structures. Since a perceptual component is included, it is not the case that representations in the mind emerge exclusively in the brain. A perceptual component is at play, as humans retrieve specific information from the actual state of the world in order to divide events, natural kinds and objects into useful categories.

Crucially, the significance of similarity diminishes in the process of categorization. This is because being similar inherently implies being different in numerous ways, across multiple dimensions, which reduces clarity on the nature of category assignment. Instead, the three pivotal components linking similar events, entities, and objects in the mind of a conceptualizer are *essences, beliefs,* and *relations* (see Section 2). For instance, in the statement "the birds chirp in the morning," the prototype bird, shaped by our experiences with bird essences encompassing size, form, and behavior, becomes linked to a specific action (chirping as opposed to, for example, bleating).

Essence assignation can be modeled by the identification of distinct features through theoretical formulation. For example, as reported by Libby et al. (2012, p. 181), the idea that theropod dinosaurs used their tails as "dynamic stabilizers during rapid or irregular movements, [has] contribut[ed] to their depiction as active and agile predators". This essentialist reconstruction of movement and stabilization profits from a shift toward a view of animal locomotion as a source of information for the development of robotic animals that emulate biological optimization of movement. In other words, a system must be open and fluctuate from orderly to near-disorderly states (critical states, see, for example, Cappelletti et al., 2021). This dynamical criticality principle can be applied as a formula to the design of an inertial appendage control for robotic stabilization (according to Libby et al., 2012). As a result we get;

$$P\ (\text{essence}) \rightarrow (\text{Verb+Space-Time Frame}_{\text{EVENT}}$$

where

- P represents the unspecified essence associated with the robotic control mechanism,
- essence pertains to the known control mechanism utilizing proportional–derivative (PD) feedback control for stabilizing body angle,

- Verb+Space-Time FrameVerb+Space-Time Frame signifies the action and spatiotemporal context of the robotic locomotion and stabilization, and
- EVENT encapsulates the overall occurrence involving the inertial appendage control of the robotic system.

This shows that a distinct continuity between organic bodily structure and the interpretation of it through active biomorphism can create semiotic definitions rooted in essentialist reconstructions. This holds paramount importance for the classification of concepts based on

Essences reflecting ecomorphological effects that generate a set of properties.

Beliefs encapsulating theories that capture significant relations encoded in natural kinds.

Relations governing the extension/exclusion of properties to/from other natural kinds.

This can be construed as an extended mechanism for essence recognition that binds living organisms on the basis of shared perceptual knowledge. An even more critical inference is that the true essence of categorization lies in embodied mechanisms influencing natural kinds and defining their characteristics. In this context, it is erroneous to presume that categorization originates from similarity by adopting a mentalistic approach aimed at creating "more theoretical treatments, which in turn could affect the details of how individual things are classified" (Hahn and Ramscar, 2011: 268).

3.1. The *Biorobotic Blending Essence Coefficient*

Artificial essence ascription proceeds from the blending of biological and artificial essence assignation. In the first place, a biomimetic robot animal is perceived as possessing distinct characteristics associated with biological referents (expanded concepts, see subsection 1.1) that may create semiotic expansion through biomorphically savvy design and/or purposeful assignation of biological features to accomplish specific tasks (involving the principle of non-local ontological dependency introduced in Section 2). Secondly, a robot animal (sign) is perceived as an artificial agent with enhanced affordances that expand the spectrum of action of biological referents to both natural and non-natural ecosystems.

The concrete relation between artificial and biological essence assignation to robot animal signs can be defined by way of the *Biorobotic Blending Essence Coefficient* (BBEC):

Robot Animal= Biological Essence+Artificial Essence/ Situated-Embedded Biomorphic-Bioinspired prototype

This formulation takes into consideration the interplay between biological and artificial essence assignation (through multiple dimensions leading to the positing of an emergent semiotic relation that integrates biological essence assignation) for the design of a robot sign, and the translation of these essences into artificial entity design:

Biological Referent Characteristics:

A robot animal (*RA*) is conceived as an entity embodying biomorphically sophisticated design (*BD*) and the deliberate allocation of biological features (that is *BF*), which "makes it explicit that the mapping to be explained is how to produce the right motor output for the given input and context" (Webb, 2008, p. 24):

$$RA=\alpha_{BD}\cdot BD+\beta_{BF}\cdot BF$$

Here, α_{BD} and β_{BF} denote the respective weighting coefficients for the biomorphic design and biological features, allowing for a nuanced adjustment of their influence in terms of functional criteria.

Artificial Agency with Expanded Affordances:

The robot animal (*RA*) becomes an artificial agent (*AA*) with enriched capabilities stemming from intentional infusion of artificial attributes (*AF*) such as learning capabilities:

$$AA=\gamma_{AF}\cdot AF$$

The parameter γ_{AF} refers to the weight assigned to the artificial attributes that permit fine-tuned control over their impact (for example, robots using proprioceptive feedback to negotiate granular terrain).

***Biorobotic Blending Essence Coefficient* (BBEC):**

The complex relationship between artificial and biological essence assignation to robot animals is quantified through the *BBEC*, considering the nuanced interplay within the *Situated-Embedded Biomorphic-Bioinspired Prototype* (SBP):

$$BBEC=\delta_{RA}\cdot RA/\chi_{SBP}\cdot SBP$$

Here, δ_{RA} and χ_{SBP} represent weighting factors, offering a granular assessment of the contribution of robot animal essence (*RA*) and the broader contextual prototype (*SBP*).

This model brings to the fore the synthesis of blended essences by offering insight into the dynamics of essence assignation to robot animal signs. The use of weighting coefficients facilitates a nuanced exploration of the intricate interplay between biomorphic design, biological features, and artificial attributes. Concurrently, the BBEC metric provides a quantitative means to evaluate the amalgamation of these essences within the expansive framework of situated-embedded prototypes. The model substantiates that signs exhibited by robot animals combine specific affordances derived from living animals, resulting in the creation of augmented agents adept at resolving targeted challenges.

It is noteworthy that the modeling of robot signs should incorporate specific ecological information related to niche construction. This underscores that, while the evolutionary trajectory of robot signs has been shaped by user-defined role assignments, primarily based on the aggregation of affordances for task completion, the next generation of robot signs will need to incorporate a multi-degree of freedom and multi-functional paradigm. The proposed paradigm integrates perception, kinetics, and neural intelligent control, simulating a diverse array of animal behaviors and successfully executing a myriad of complex tasks (Gao et al., 2019, p. 348). Seen in this light, the robot animal sign can be defined in terms of the formula:

Robot Animal Sign = representation [(icon+index+symbol^n) x (essence assignation+ affordance reconstruction)]/machinic semiosis

Let's break down the semiotic model for robot animal signs using the provided formula, let's define each component and introduce quadratic equations if necessary:

Representation (R): The manner in which the robot animal sign is visually or conceptually depicted.

Quadratic Equation: $R=a_1x^2+b_1x+c1$

Icon (I): A visual representation that evokes a direct connection to the perceived object or concept.

Quadratic Equation: $I=a_2x^2+b_2x+c^2$

Index (X): A sign that points to specific features associated with the robot animal in terms of reaction to environmental conditions.

Quadratic Equation: $X=a_3x^2+b_3x+c_3$

Symbol (S_n): A complex sign incorporating various attributes, where $_n$ denotes the multiplicity of interpretative laws associated with context-bound application/function of a robot sign.

Quadratic Equation: $S_n=a_4x^2+b_4x+c_4$

Essence Assignation (EA): The attribution of non-observable qualities such as "fishness" or "spiderness" to a robot animal.

Quadratic Equation: $EA=a_5x^2+b_5x+c_5$

Affordance Reconstruction (AR): The process of assigning affordance-driven performance to the robot animal sign for problem-solving tasks.

Quadratic Equation: $AR=a_6x2+b_6x+c_6$

Machinic Semiosis (MS): The dynamic interpretative process within the machinic framework.

Quadratic Equation: $MS=a_7x^2+b_7x+c_7$

The current model seeks to provide an epistemology for the introduction of more ecological representations of the multilayered interplay of animal iconicity, behavioral symbolicity, human essence assignation to animal signs, and the ensuing relocation of robot animal signs as potential autonomous agents. The model also provides a phenomenological reading of the potential emergence of robot consciousness, which also delves into complex questions regarding the role of autonomy in the biological representation of the world. The idea that robot animal signs may at some point of their technological evolution respond to questions such as "what I do next?" (Franklin, 1995), also pushes the boundaries of what we consider as a living entity. Animals are generally defined as such on the basis of animacy or, as is the case of robot animals, in terms of the conjunction of seeming animacy and purpose engineered by humans. The idea of technological myth begins precisely with the evidence that the narratives serving as conceptual antecedents for the development of a robot sign are, essentially, techno fairy tales in which a prototypical animal agent *comes to life* in order to bring some idea of progress to an end. Thus, the whole process of modeling a potentially usable robotic animal encompasses the tacit assumption that this sign will be part of an extended semiosis embedded in the human interpretation of animal action-decision loops. Even with the advent of AI, the necessary adjustment of the artificial agent to their environment is still regulated by human-centered paradigms that ignore biological imperatives such as self-replication for continuity in an environment, which has raised a number of questions regarding the moral standing of robot animals (e.g., Coeckelbergh, 2021; Coghlan et al., 2019; Gordon, 2018) and whether robots are eligible for rights (Gellers, 2021). Phylogenetic adaptation is thus replaced by training models and actuator refinement (see Matharu et al. 2023). So, it is clear that movement emulation and biomorphic precision need to incorporate a suitable model of internal states that guarantee the preservation of robot signs as autonomous agents in non-controlled environments. This means, among other things that the robot animal sign's symbolic contour needs to be defined independently from function. Although animals are not robots (Schilhab, 2013), this does not preclude that robots be perceived as artificial animals. Furthermore, the claim that robot animal (signs) is a construct that only serves the purpose of "upgrading the cognitive powers of machines" (Schilhab, 2013, p. 2) is a proof to the cognitive bias and semiotic blindness toward sentiency that still plague our ideas about non-human and non-organic entities. Indeed, there is mounting evidence which suggests that artificial intelligence may at some point push the boundaries of intelligent animacy and acquire an ontological status well beyond essentialist definitions.

4. Conclusion

The exploration of enhanced biomorphic qualities in robot

design reveals a complex interplay between biological referents and artificial affordances. In this context, the emergence of biomimetic robots, characterized by reconstructions that emulate biological features, raises fundamental questions about the nature of animacy and the semiotic relationship between artificial and biological entities. The suggestion of continuity between artificial and biological entities challenges the notion that animals are devoid of sentience, presenting robots as models for understanding organism behavior. However, the biomorphic connection between robot animal signs and their biological counterparts introduces a semiotic gap, which questions whether robot animals are true semiotic extensions or mere diagrams lacking internal values. The deployment of specific affordances, inspired by trial-and-error learning algorithms in response to injuries, emphasizes the potential adaptability and learning capabilities of non-animal species. This advances us far toward an emerging new shift in perspective, whereby AI-powered robot animal signs do not merely simulate sentience but have the potential to develop non-human forms of intelligence. It goes without saying that a proper treatment of this topic would obviously require considerably more space than I can afford to devote to it here. However, the proposed model offers an epistemology for ecological representations that integrates the interplay of animal iconicity, behavioral symbolicity, and human essence assignment to robot animal signs. Although these remarks raise a number of different matters, the possibility of robot consciousness as a result of the existence internal states that are meaningful to the machines, open avenues for the optimization of strategies in niche construction. As we explore the evolving conceptual landscape of AI-powered robot animals, it becomes apparent that human-centered paradigms still regulate their adjustment to the environment in the belief that "life-like" *in silico* artificial animacy, "requires performative, not textural, simulation" (Helmreich, 2007, p.330). The reliance on training models and actuator refinement, rather than phylogenetic adaptation, raises ethical concerns about the moral standing of robot animals. The need for a suitable model of internal states, independent of function, underscores the importance of defining the symbolic contour of robot animal signs in non-controlled environments. In essence, while animals are not robots, the assertion that robot animals solely serve the purpose of upgrading machine cognitive powers reflects cognitive bias and semiotic blindness. The evidence suggests that artificial intelligence has the potential to redefine intelligent animacy beyond essentialist definitions, thereby challenging our perceptions of non-human and non-organic entities. This opens new avenues for understanding and embracing cognitive diversity in the expanding realm of artificial intelligence.

References

Cappelletti, G., D'Addese, G., Serra,R., & Villani, M. (2021). Dynamical criticality in growing networks. In Schneider, J.J, Weyland, M.S., Flumini, D. & Füchslin, R.M. (Eds.), *Artificial Life and Evolutionary Computation. 15th Italian Workshop, WIVACE 2021, Winterthur, Switzerland, September 15–17, 2021, Revised Selected Chapters*, pp. 3-13. Cham: Springer Nature Switzerland AG.

Coeckelbergh, M. (2011). Humans, animals, and robots: A phenomenological approach to human-robot relations. *International Journal of Social Robotics, 3,*

197-204. https://doi.org/10.1007/s12369-010-0075-6

Coeckelbergh, M. (2021). Should we treat teddy bear 2.0 as a Kantian dog? Four arguments for the indirect moral standing of personal social robots, with implications for thinking about animals and humans. *Minds and Machines, 31*, 337–360. https://doi.org/10.1007/s11023-020-09554-3

Coghlan, S., Vetere, F., Waycott, J., & Barbosa Neves, B. (2019). Could social robots make us kinder or crueller to humans and animals? *International Journal of Social Robotics*. https://doi.org/10.1007/s12369-019-00583-2

Christiansen, M. B., Rafsanjani, A., & Jørgensen, J. (2023). "It brings the good vibes": Exploring biomorphic aesthetics in the design of soft personal robots. *International Journal of Social Robotics*. https://doi.org/10.1007/s12369-023-01037-6

Cully, A., Clune, J., Tarapore, D., & Mouret, J.-B. (2015). Robots that can adapt like animals. *Nature, 521*, 502-514.

Denby, D. (2014). Essence and Intrinsicality. In Francescotti, R. (Ed.), *Companion to intrinsic properties*, pp. 87-110. Berlin/Boston: Walter DeGruyter.

Facchin, M. (2024). Neural representations unobserved—or: a dilemma for the cognitive neuroscience revolution. *Synthese*, 203(7). https://doi.org/10.1007/s11229-023-04418-6

Favela, Luis H. & Machery, E. (2023). Investigating the concept of representation in the neural and psychological sciences. *Frontiers in* Psychology, 14, 1165622. doi: 10.3389/fpsyg.2023.1165622

Franklin, S. (1995). *Artificial minds*. Cambridge MA: MIT Press.

Gao, Z., Shi, Q., Fukuda, T., Li, C., & Huang, Q. (2019). An overview of biomimetic robots with animal behaviors. *Neurocomputing, 332*, 339–350. https://doi.org/10.1016/j.neucom.2018.12.071

Gellers, J. C. (2020). *Rights for robots: Artificial intelligence, animal and environmental law*. Oxon: Routledge.

Gibson, J. J. 1969. *The senses considered as perceptual systems*. Boston: Houghton Mifflin.

Gibson, J. J. 1977. *The ecological approach to visual perception*. Boston: Houghton Mifflin.

Gordon, J.-S. (2018). What do we owe to intelligent robots? AI & SOCIETY. https://doi.org/10.1007/s00146-018-0844-6

Helmreich, S. (2007). An archaeology of artificial life, underwater. In Riskin, J. (Ed.), *Genesis Redux, essays in the history and philosophy of artificial life*, pp. 321-333. Chicago and London: University of Chicago Press.

Hipólito, I. (2022). Cognition without neural representation: dynamics of a complex system. *Frontiers in Psychology 12, 643276*. doi: 10.3389/fpsyg.2021.643276

Korondi, P., Korcsok, B., Kovács, S., & Niitsuma, M. (2015). Etho-robotics: What kind of behaviour can we learn from the animals? *IFAC-ChaptersOnLine, 48*(19), 244–255. https://doi.org/10.1016/j.ifacol.2015.12.041

Kurokawa, S. (2022). Evolution of trustfulness in the case where resources for cooperation are sometimes absent. *Theoretical Population Biology, 145*, 63–79. https://doi.org/10.1016/j.tpb.2022.03.002

Libby, T., Moore, T. Y., Chang-Siu, E., Li, D., Cohen, D. J., Jusufi, A., & Full, R. J. (2012). Tail-assisted pitch control in lizards, robots, and dinosaurs. *Nature, 481*, 181–184. https://doi.org/10.1038/nature10710

Lewis, D. (1983). Extrinsic properties, *Philosophical Studies*, 44, 197–200.

Long, J. (2012). *Darwin's Devices: What evolving robots can teach us about the history of life and the future of technology*. New York: Basic Books.

Malt, B.C. (2024). Representing the world in language and thought. *Topics in Cognitive Science* 16, 6–24.

Marshall, D., & Weatherson, B. (2002). Intrinsic vs. extrinsic properties. *Stanford Encyclopedia of Philosophy*. Online available: https://plato.stanford.edu/entries/intrinsic-extrinsic/

Odling-Smee, F. J., Laland, K. N., & Feldman, M. W. (2003). *Niche construction: The neglected process in evolution*. Princeton, MA: Princeton University Press.

Parisi, D. (2014). One artefact, many phenomena. In Cagnoni, S., Mirolli, M., & Villani, M. (Eds.), *Evolution, complexity and artificial life*, pp. 3-18.

Cham: Springer Nature AG.

Peng, Y., Wang, Z., Zhang, Q., Du, S., Zhao, Y., Yang, L., Liu, J., Cheng, Y., Wang, A., & Liu, Y. (2019). Basic research on wireless remote control rabbit animal robot movement. In H. Yu, J. Liu, L. Liu, Z. Ju, Y. Liu, & D. Zhou (Eds.), *Intelligent Robotics and Applications. ICIRA 2019. Lecture Notes in Computer Science, vol. 11741*. Springer, Cham. https://doi.org/10.1007/978-3-030-27532-7_4

Romano, D., Donati, E., Benelli, G., & Stefanini, C. (2018). A review on animal–robot interaction: From bio-hybrid organisms to mixed societies. *Biological Cybernetics*. https://doi.org/10.1007/s00422-018-0787-5

Schilhab, T. S. S. (2013). Why animals are not robots. *Phenomenology and the Cognitive Sciences, 14*, 599–611. https://doi.org/10.1007/s11097-013-9342-y

Stanton, C. M., Kahn, J. P. H., Severson, R. L., Ruckert, J. H., & Gill, B. T. (2008). Robotic animals might aid in the social development of children with autism. In *ACM HRI 2008*, Amsterdam, Netherlands, 12-15.

Vallentyne, P. (2014. Intrinsic properties defined. In Francescotti, R. (Ed.), *Companion to intrinsic properties*, pp. 31-40. Berlin/Boston: Walter DeGruyter.

von Uexküll, J. (1982). The theory of meaning. *Semiotica, 42* (1), 25-82. [Translation of 'Bedeutungslehre' (U. 1940a), with a preface, introductory chapter, and glossary by the guest editor of the issue. Thure von Uexküll (1982a, 1982b, 1982c). Trans. by Barry Stone and Herbert Weiner.]

Wang, J., Chen, W., Xiao, X., Xu, Y., Li, C., Jia, X., & Meng, M. Q.-H. (2021). A survey of the development of biomimetic intelligence and robotics. *Biomimetic Intelligence and Robotics*. https://doi.org/10.1016/j.birob.2021.100001

Webb, B. (2008). Using robots to understand animal behavior. *Advances in the Study of Behavior, Academic Press, 38*, 1-58. doi.org/10.1016/S0065-3454(08)00001-6

Wilson, M. (2023) *Imitation of rigor: An alternative history of analytic philosophy*. Oxford: Oxford University Press.

Zhou, Z., Mei, H., Li, R., Wang, C., Fang, K., Wang, W., Tang, Y., & Dai, Z. (2022). Progresses of animal robots: A historical review and perspectiveness. *Heliyon, 8*, e11499. https://doi.org/10.1016/j.heliyon.2022.e11499

INDEX

R

S

T

U

W

HIGH-RESOLUTION IMAGES

Chapter 1, Table 1.1. Frequencies of modal constructions in the *Corpus of Contemporary American English*.

Modal construction	TYPE	FREQ	PER MIL	FORM	FUNCTION
COULD BE (+NP)	Defining MRC	21,685	198.23	(**Subj.$_{SOURCE}$**) **MODAL DO/BE NP**	**[X MODAL DO/BE Y$_{DEF}$]**
COULD SEE	Transitive MASC	17,145	163.44	(**Subj.$_{FORCE}$**) **MODAL V Obj.**	**[X MOD act on/experience [Y]$_{SCOPE}$**
CAN GET	Caused-motion MASC	10,864	99.31	(**Subj.$_{FORCE}$**) **MODAL V [Oblique Source]** $_{SCOPE}$	**[X MOD cause [Y to move from Z]$_{SCOPE}$**
WILL GET	Intransitive-motion	9,509	86.93	(**Subj.$_{FORCE}$**) **MODAL V [Oblique$_{PATH}$]** $_{SCOPE}$	**[X MOD move [Z$_{PATH}$]$_{SCOPE}$**
WILL MAKE	Ditransitive MASC	4,232	38.43	**Subj.$_{FORCE}$) MODAL V [Obj1 Obj2]** $_{SCOPE}$	**[X MOD cause [Y to receive Z]$_{SCOPE}$**
WILL BE ABLE	Attributive MRC	3,039	27.78	**Subj.$_{SOURCE}$) MODAL BE Adj (to phrase)**	**[X MODAL BE Y$_{STATE}$]**

Modal construction	TYPE	FREQ	PER MIL	FORM	FUNCTION
COULD BE (+NP)	Defining MRC	21,685	198.23	(**Subj.$_{SOURCE}$**) **MODAL DO/BE NP**	**[X MODAL DO/BE Y$_{DEF}$]**
COULD SEE	Transitive MASC	17,145	163.44	(**Subj.$_{FORCE}$**) **MODAL V Obj.**	**[X MOD act on/experience [Y]$_{SCOPE}$**
CAN GET	Caused-motion MASC	10,864	99.31	(**Subj.$_{FORCE}$**) **MODAL V [Oblique Source]** $_{SCOPE}$	**[X MOD cause [Y to move from Z]$_{SCOPE}$**
WILL GET	Intransitive-motion	9,509	86.93	(**Subj.$_{FORCE}$**) **MODAL V [Oblique$_{PATH}$]** $_{SCOPE}$	**[X MOD move [Z$_{PATH}$]$_{SCOPE}$**
WILL MAKE	Ditransitive MASC	4,232	38.43	**Subj.$_{FORCE}$) MODAL V [Obj1 Obj2] SCOPE**	**[X MOD cause [Y to receive Z]$_{SCOPE}$**
WILL BE ABLE	Attributive MRC	3,039	27.78	**Subj.$_{SOURCE}$) MODAL BE Adj (to phrase)**	**[X MODAL BE Y$_{STATE}$]**

Chapter 1, **Table 1.2.** Frequency of *"be able to"* in the *Corpus of Contemporary American English.*

SECTION	ALL	BLOG	WEB	TV/M	SPOK	FIC	MAG	NEWS	ACAD	1990-94	1995-99	2000-04	2005-09	2010-14	2015-19
FREQ	23145	0	0	0	23145	0	0	0	0	4053	3840	3979	3666	3909	3698
WORDS (M)	993	21.8	21.1	21.8	21.5	20.1	21.5	20.6	20.3	23.7	25.1	24.9	24.5	24.7	24.5
PER MIL	23.31	0	0	0	1,076.73	0	0	0	0	171.03	152.48	159.3	148.41	157.91	149.92

Chapter 2, **Figure 2.3**. The cycle of language generation of ChatGPT.

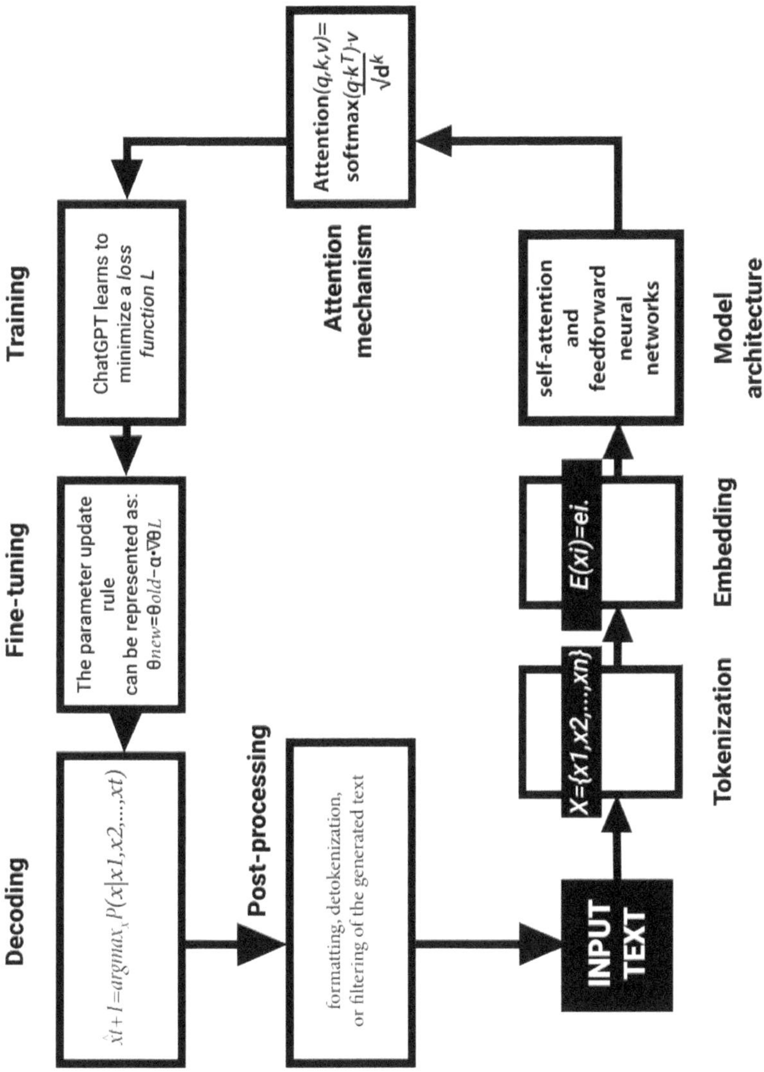

About the author

Sergio Torres–Martínez is professor of cognitive linguistics, semiotics and translation semiotics. Among his main interests are *Agentive Cognitive Construction Grammar*, Cognitive Semantics, embodiment theory, phenomenology, Wittgenstein's philosophy of language, Peircean semiotics and the cognitive applications of construction grammar (*Applied Cognitive Construction Grammar*). Current research projects include the conceptualization of construction grammar as an interdisciplinary field of endeavor connecting embodiment theory, neuroscience semiotics and philosophy for the construction of a comprehensive and systematic description of constructional attachment patterns across languages. Central to this research is the need to provide linguistics with a model of the mind that complements linguistic description.

Thank you for purchasing this *MÍMIR Editors* book.
Gracias por comprar este libro publicado por *MÍMIR Editors*.

MÍMIR books may be purchased for business or promotional use.
For information on bulk purchases please contact the editor at

mimireditors@gmail.com

Los libros de MÍMIR Editors pueden adquirirse para fines comerciales o promocionales. Para información sobre ventas al por mayor, por favor contacte al editor

mimireditors@gmail.com

www.ingramcontent.com/pod-product-compliance
Ingram Content Group UK Ltd.
Pitfield, Milton Keynes, MK11 3LW, UK
UKHW040228290926
15471UKWH00001B/60

9 798230 639725